Future-Proofing Industries: AI-Driven Security, ERP, and Risk Management

Ashish Joon
Harshit Kohli

Made with ♥ on the Notion Press Platform

www.notionpress.com

Dedications

Ashish Joon

To those who taught me that growth often begins in silence, and strength is built in moments unseen.

To every obstacle that became a lesson, and every lesson that became a stepping stone.

To the people—family, friends, mentors—whose presence reminded me that I was never walking this path alone.

This book is for the seekers of meaning, the ones who chase not only success, but purpose. May you always find light in the journey itself.

Harshit Kohli

The many people who find themselves at the pivotal intersection of technology and humanity in the healthcare industry are honoured in this book.

To the committed doctors, nurses, and clinical professionals who embrace change while maintaining their commitment to compassion, who adjust to new systems without breaking their core promise to heal, and who view artificial intelligence as a helpful ally rather than a replacement in their admirable pursuits.

To the data scientists, engineers, and AI researchers who apply their knowledge to tackle the most difficult problems in healthcare, who develop code with a deep understanding of the people their algorithms may affect, and who maintain an unwavering commitment to moral advancement. This commitment also includes the patients whose confidence drives progress, whose information aids in education, and whose stories serve as a moving reminder of the important reason for each medical technology breakthrough.

Furthermore, to the regulators, legislators, and healthcare administrators who negotiate difficult obstacles to guarantee that AI systems fairly benefit everyone while upholding their right to privacy and dignity. To the educators who are teaching both technical skills and humanistic values to prepare the workforce of the future for an artificial intelligence-transformed healthcare environment. Last but not least, to the pioneers—both well-known and unsung heroes—who dared to envision a time where human wisdom and artificial intelligence coexist peacefully to expand the possibilities for healing, lessen suffering, and recognize the inherent value of every person.

With a constant focus on the welfare of humanity, it is hoped that this book will contribute to the careful integration of AI into the complex art and science of medicine.

Contents

Preface

In an era defined by rapid digital transformation, artificial intelligence is no longer a futuristic concept—it is the present force redefining how industries operate, secure their systems, and manage risk. Future-Proofing Industries: AI-Driven Security, ERP, and Risk Management was conceived to address this shift, offering a comprehensive guide to the ways AI is revolutionizing core enterprise functions.

This book explores the integration of AI in cybersecurity, enterprise resource planning, financial fraud detection, and strategic risk management. It draws from both current industry practices and emerging trends to provide a future-facing perspective for leaders, technologists, and researchers.

The motivation behind this book is simple: to provide a resource that bridges theory and application, empowering organizations to harness AI not only to solve today's problems but also to build resilience for tomorrow.

Each chapter is designed to offer actionable insights, critical evaluations, and thought leadership for those seeking to navigate the complexities of AI adoption responsibly and effectively.

We hope this book serves as a valuable guide and a catalyst for strategic innovation.

Acknowledgment

This book would not have been possible without the invaluable support, guidance, and encouragement of many individuals and institutions.

First and foremost, we would like to express our deepest gratitude to our mentors and colleagues, whose insights and constructive critiques have shaped the direction and depth of this work. Your dedication to the evolving field of artificial intelligence continues to inspire.

We are sincerely thankful to the academic and professional communities whose research and contributions have laid the foundation for this book. Your work continues to push the boundaries of innovation in security, ERP, and risk management.

To our family and friends, your unwavering support has been a constant source of strength. Thank you for your patience and encouragement throughout the writing process.

Lastly, we extend our appreciation to readers and practitioners seeking to embrace AI's potential responsibly and strategically. This book is written for your curiosity, ambition, and forward-thinking mindset.

Chapter 01

AI in Security and Threat Prevention

1.1 Introduction to AI in Security

Artificial Intelligence (AI) has been a game changer when it comes to preparing the security field for automation, intelligence, and adaptation of automatic mechanisms that are able to detect, prevent, and respond to threats. As digital infrastructure and, therefore, systems interconnect, the associated threat surface rapidly grows, and traditional security measures are insufficient for responding to attacks and other types of threats of an advanced nature. The AI-driven security solutions based on machine learning, deep learning, and advanced analytics help better protect the risk management, intrusion detection as well as threat mitigation. Real time, these intelligent system process massive amounts of information to detect patterns and anomalies that would suggest, or possibly have already taken place, a security breach or a process that is malicious. With security threats evolving, AI is increasingly becoming a part of strengthening the defense mechanisms in different industries (Horowitz et al., 2018).

Cybersecurity is one of the most important aspects where AI is revolutionizing security. Geared towards power, AI can analyse network traffic, spots vulnerabilities, and fight back cyberattacks without much of human interference. Due to their highest level of effectiveness, machine learning algorithms are perfect for finding malicious patterns, thanks to which the security teams can proactively defend against ransomware, phishing, and some APTs. Furthermore,

AI-based threat intelligence systems always search the past security incidents to learn and develop their capacity to identify the new attack vector. But this level of automation not just cuts response time but also helps lessens the chances of human error, which thoroughly increases the robustness and resilience of cybersecurity.

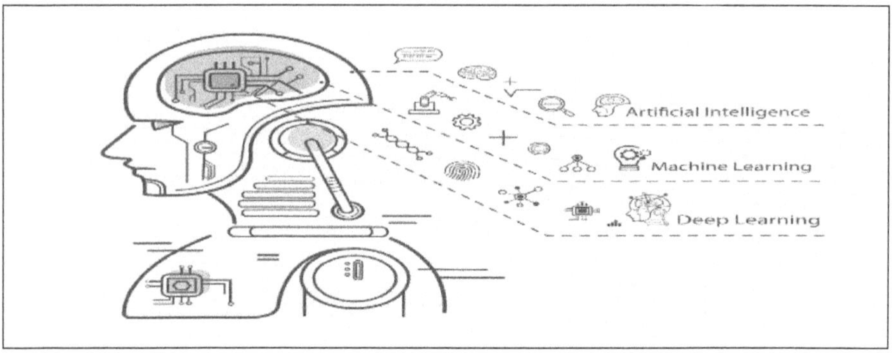

Figure 1.1 *AI, Machine Learning, and Deep Learning Conceptual Diagram.*

Source: (Seker, 2020)

AI is also changing a physical security through advancements in surveillance and the biometric authentication. Facial recognition, object detection, behavioral analysis, are used techniques by AI driven surveillance system to monitor image stream and detect suspicious activities in real time. Traditional security measures are unable to accurately identify the potential threats as compared to these systems which can identify the potential threats better so the chance of a security breach is less. Moreover, various AI-enhanced biometric authentication methods such as fingerprint, voice, and iris recognition systems have become very secure to prevent unauthorized people from accessing areas with high security. The particular value of AI as a contributor to physical security is in high-risk environments including airports, government facilities and financial institutions.

Risk management and regulatory compliance, which are another important role of AI in security, can be viewed as another critical application of AI in security. To ensure a bank follows stringent security regulations and complies with high compliance frameworks in order to protect sensitive data and avoid financial frauds. These AI driven compliance monitoring systems update the business with the latest regulatory requirements and looks for potential compliance violations on large datasets. The Natural Language Processing (NLP) algorithms help parse legal documents and security policies for the organizations to maintain the standard of industry. Furthermore, AI based risk assessment can predict the potential security vulnerability before such security is exploited by organizations to take preventive actions proactively (Walters & Novak, 2021) China (Hong Kong, Macau, Taiwan.

It's true that AI has a number of advantages in security, but it also poses some challenges and issues regarding ethics. The concern for the biggest bias in AI models remains one of the main ones to take up. This means that if AI security systems are trained on biased datasets, then such systems may create false threat assessments and ultimately racially discriminate against certain individual. Moreover, the use of AI driven surveillance and biometric authentication give raises to at least the same types of privacy concerns as these technologies collect and process huge amounts of personal data. Also, cyber attackers are doing adversarial attacks against AI systems in which the malicious actors feed misleading or manipulated data to the AI models. To solve these problems, strict governance, transparency and constant model improvement are needed to ensure that AI security solution will stay effective and moral.

Finally, it stands to say that AI is about to reshape security through the use of advanced automation, predictive analytics, as well as real time threat detection. It is used in cybersecurity, physical security, and risk management to bolster the ability of organizations to develop more resiliency in their underlying defense mechanisms against ever changing threats. But if you want to successfully integrate AI into security, ethics and the regulatory compliance, plus vulnerabilities possible in the adoption of the AI in the security process need to be taken into account. With AI technology evolving continuously, there will be more and more of an imperative for its role in security and more and more sophisticated and intelligent security solutions will come as a result.

1.2 AI-Driven Cybersecurity Frameworks

Artificial intelligence changes the way that organizations detect, respond and prevent threats, and it has integrated with cybersecurity. AI driven cybersecurity framework is used to managing security functions by using always on advanced technologies of machine learning (ML), deep learning (DL), natural language processing (NLP) and automation. The framework consists of layers that are addressing critical issues of the cybersecurity. The comele of AI driven security mechanisms helps organizations to mitigate the risk of cyber threats, lower their response time and strengthen their defence against the cyber threats. The rest of this section introduces the main elements of an AI enabled cybersecurity architecture (Parisi, 2019).

1. Threat Detection & Prevention

The real time threat detection and mitigation systems based on AI are helpful to an organization to detect and take up the cyber threats better at the time of occurrence. For example, antivirus programs and intrusion detection systems based on signatures and rules are not able to keep up with the high speed of cyberattacks. But AI powered solutions are automated solutions that can analyze huge amounts of network traffic and can identify and detect abnormality as well as the malicious behavior of the network accurately.

Behavioral analysis is one of the key elements of the AI threat detection. The network patterns and user activity are constantly being learned by the AI models so that they might differentiate normal and suspicious behaviour. If an AI system detects it, it can flag an unusual login attempt, file access patterns, unauthorized data transfers as possible security threats. In improving the malware detection, machine learning algorithms also help by learning and classifying and identifying the malware invariants by their behaviour rather than relying on the predefined signatures. For one, phishing detection is also another area where AI is used to do — natural language processing checks emails, messages and websites for signs of phishing emails to reduce the risks of lost credentials and fraud.

2. Automated Incident Response

Automation of incident response is crucial to cut down time spent in responding to cyber threats and damage caused. Security Orchestration, Automation and Response (SOAR) with AI capabilities for security empower organizations to manually put security tasks that are repetitive in play and further improve incident management. These platforms help in checking the security alerts, deciding the importance of threats and triggering pre-set action responses to fight off the security incidents before they run out of control.

The use of AI for scoring risk further enriches the incident response functionality by enabling the determination of severity of the threat based on the intensity and the probability of exploitation. By focusing security team's limited resources on top ranked threat instead of spending their time chasing bogus points, this improves security team's ability to react to important threats without becoming bogged down with the lesser ones. Additionally, the self-healing security system based on AI automatically patches the vulnerable software, isolates the system and restore the affected services without any human intervention. Integrating security into automation reduces the arms race of response time, reduces human error, and increases security efficiency.

3. AI-Enabled Threat Intelligence

Threat intelligence is one of the basic parts of modern cybersecurity framework, making organizations able to stay ahead of the flank of newer cyber threats. With AI, threat intelligence collects and analyzes huge amounts of security data from different sources, i.e. threat reports, dark web forums and social media. NLP allows AI to be able to process unstructured data and to extract key insights regarding potential cyber threats. It gives the security teams a chance to reactively stop new attack vectors before they spread widely.

With historical attack data and the recognition of their tendencies, machine learning models help in better recognition of attack patterns. This predictive capability allows them to gain an understanding of how cybercriminals operate to counter advanced persistent threats (APTs). In addition to this, AI powered threat hunting tools also proactively look for hidden threats within the network of an organization and will hunt for stealthy attacks that traditional security solutions might miss. Utilizing AI for threat intelligence allows organizations to develop an offensive defense strategy and become much more responsive in the face of security threats as they emerge.

4. Identity & Access Management (IAM)

The first one is that cybersecurity ensures secure access to digital resources. Identity and Access Management (IAM) using AI is a hot topic as it can address the two critical issues in security biometrics authentication and continuous monitoring with adaptive access controls. In contrast, AI driven biometric authentication methods like facial recognition, fingerprints scanning or voice recognition are more secure alternative to the password based one; they are less vulnerable to the theft of credentials.

Adaptive authentication changes the risk assessment dynamically on a recurring basis taking into consideration the device type, location and user behavior to check if extra authentication steps are needed. To illustrate, if an employee tries to get to sensitive data from a new location, then AI can invoke multi factor authentication (MFA) or won't allow access unless further verification is done. The Zero Trust

Security Model when it comes with the principle of 'never trust, always verify' is also further enhanced by AI. This creates a situation where every user and device trying to get to an organization network is constantly being authenticated and authorized, reducing the chance of unauthorized access and insider threats.

5. Data Protection & Privacy Enhancement

In all sorts of industries, organizations are on the lookout for protecting sensitive data from cyber threats. AI driven encryption techniques make the data more secure by automatically encrypting the sensitive data in it and allow only authorized users to access it. The ability of AI based on encryption algorithms to develop itself against cryptographic attacks is more complicated for cybercriminals to break into data integrity.

Differential privacy and Federated learning techniques make AI to be privacy preserving techniques of learning from the user data without compromising the privacy of the user. To achieve so, it adds noise, or noise, to datasets that make user information harder to infer, but still guarantee accurate AI predictions. Provided that the data does not require access to the raw user data, federated learning permits AI models to be trained in the absence of raw user data, as long as privacy can be maintained and security is upheld.

Data protection also involves being compliant with regulatory compliance. Compliance monitoring with the help of AI requires organizations to continuously scan data-handling practices, alerting if any breaches might occur in such cases. These regulations include the General Data Protection Regulation (GDPR), Health Insurance Portability and Accountability Act (HIPAA), and California Consumer Privacy Act (CCPA). Organizations can let AI influence the confidentiality at the data, protect their data from being breached, and adhere to international data security standards by incorporating AI in their data security strategy.

1.2.1 AI for intrusion detection and prevention

Industrial environments are increasingly tied in to the world of the internet, which means that we are experiencing an increase right now in cyberattacks of all types on industrial environments. To this end, Wireless Sensor Networks (WSNs) have been incorporated to monitor and automate in real time as a part of Industry 4.0. However, these networks are quite vulnerable to blackhole attacks, gray hole attacks, flooding attacks and scheduling attacks. To counter these problems, Artificial Intelligence (AI) offers a strong tool for intrusion detection and prevention in industrial systems to boost the level of cybersecurity resilience.

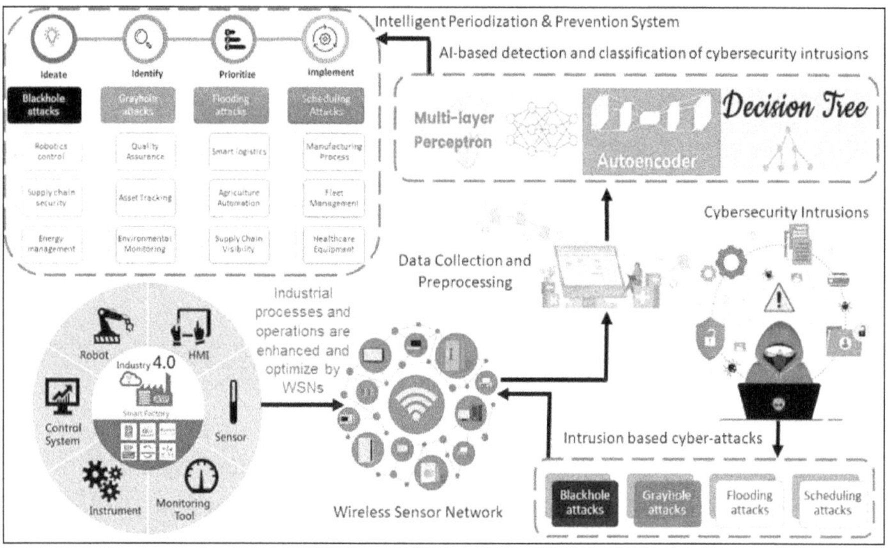

Figure 1.2 *Cybersecurity Intrusion Detection System in Industrial Wireless Sensor Networks (WSNs).*

***Source:** (Al-Quayed et al., 2024)*

Machine learning AI Intrusion Detection Systems on the other hand are using them, to analyze network traffic and identify anomalies such as anomalies or threat. There is a gap to fill, and AI-driven solutions are important to fill it because the ever-more complex modern cyber-attacks can outpace traditional cybersecurity mechanisms. These are high precision techniques in terms of threat identification, anomaly detection and automated classification, and also a highly efficient cybersecurity operations running method.

Cybersecurity can be beyond detection and involves intelligent periodization and prevention mechanisms with AI. Threats can be prioritized on the basis of priority such as severity, impact and system vulnerabilities and the organizations will be able to undertake those mitigation strategies accordingly. In other words, inclusion of AI in cybersecurity framework can assist seasoned the network of industries and protect such critical assets and operational type asecenece while being attacked increasingly sophisticated cyber-attacks.

1.2.2 AI-powered threat intelligence and risk scoring

Cybersecurity is being changed through real-time threat intelligence with dynamic risk scoring using Artificial Intelligence (AI). Due to the ever-changing cyber threat landscape, traditional security approaches cannot keep up and therefore have a minimal ability and cannot perform well against the traditional ones. It is able to tap into the power of machine learning and big data analytics to analyze huge amounts of security data, find patterns, and understand what the patterns are, and what type of pattern will turn to a problem. This thinking in terms of scheduling in advance enables organizations to make much more proactive security decisions and to react much faster to threats (known or not). Matching and crossing across data between different platforms allows AI models to detect threat across various planes in an organization's security landscape that would otherwise be missed (Parisi, 2019).

Threat intelligence consists of data from various sources i.e. network's logs, dark web monitoring, cloud environments, and endpoint activity to identify the abnormal and suspicious things. Deep learning and NLP help to improve threat classification in advanced models to enhance the risk reassessment. Automatically assigned severity levels are given to threats via risk scoring where threats are classified based on the historical attack data, behavior analysis and contextual factors for response prioritization and resource allocation. AI is able to work on large datasets and quickly identify high-risk activities making it easier through which a security team can respond effectively and mitigate potential damages.

Figure 1.3 *AI-Driven Cybersecurity: Human-Machine Collaboration.*

Source: (TARLOGIC, 2023)

AI empowered cybersecurity solutions bring up threats faster, have a lesser false positive rate and automated mitigation strategy to counter them. In addition to securing the systems, these capabilities offload the workload from human analysts to focus on high priority incidents. With AI driven systems, dynamic market threats changing over a period of time can be provided with dynamic security policies for a better protection. To better address the risks associated with cyber threats getting more and more sophisticated and scalable, it makes sense to bring the efficiency and resiliency of risk management postures forward with AI, which, we know, is learning and on the go. Leveraging AI into the threat intelligence frameworks allows enterprises to encompass more and more complex attacks and provide a solid and hard rock protection to some of the most important digital assets and industrial operations. Additionally, it also automates the proactive security culture, which is being digitally resilient around the now changing threat landscape.

1.3 Machine Learning for Malware Detection and Intrusion Response

Machine learning has revolutionized the malware detection and intrusion detection and response by automating and bringing intelligence to automated threat identification and mitigation. However, traditional signature-based antivirus systems find it difficult to address the fast pace of the emergence of cyber threats as cyber attackers are continuously creating new malware variant to breach conventional defense and compromise a host system. But much like ML powered systems, advanced pattern recognition, anomaly detection and predictive analytics tools used in malware determine cause and classify them based on behavioral characteristics as opposed to static signatures. An ML can train models on huge dataset which contain the known malicious activities and benign activities and it can identify zero-day threats, polymorphic malware, and Advanced Persistent Threat (APTs) that conventional methods could miss.

In addition, ML improves intrusion response by automating threat mitigation and decreasing response time. Security frameworks based on AI powered methods make use of supervised as well as unsupervised techniques for real time monitoring of the network traffic, detection of the anomalies, and categorization of the attack vectors. ML based systems are able to automatically respond taking the form of isolating compromised endpoints, blocking malicious IP addresses and flagging high risk activities for further investigation when intrusion attempts are observed. Using convolutional neural networks (CNNs) and recurrent neural networks (RNNs), the intrusion detection method is further deepened to detect contorted attack patterns and reduce the false positives allowing for the time of security force to direct their focus to high priority threat.

ML brings to the table stronger overall resilience, as it introduces adaptive and self-learning security systems. ML algorithms are no static rule-based security model, on the contrary ML algorithms learn (continually) from new attack patterns and adapt their detection capabilities without any manual intervention. By enhancing the accuracy of threat detection and reducing the burden from human analysts on strategic security initiatives, this dynamic approach also reduces the burden on analysts to focus on strategic security initiatives. With the rise of sophisticated cyber threats, ML driven malware detection and intrusion response will be critical to protect critical infrastructures, enterprise networks as well as digital assets against an increasing number of cyberattacks (He, 2011).

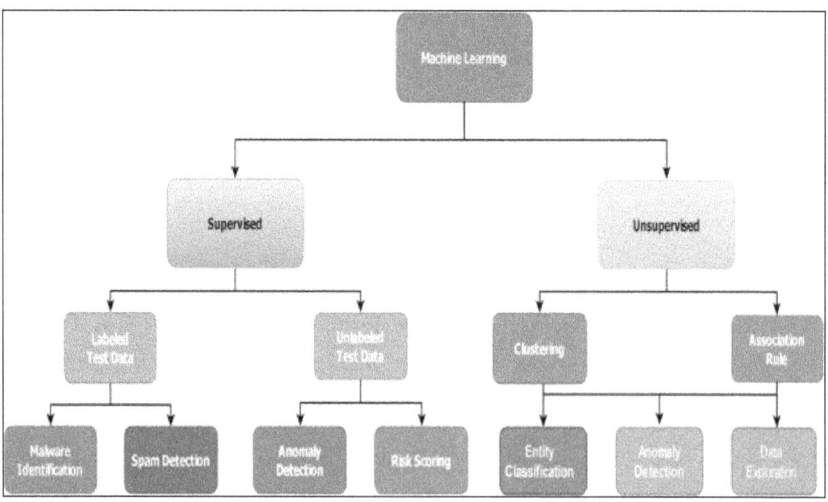

Figure 1.4 *Machine Learning Taxonomy: Supervised vs. Unsupervised Learning.*

Source: (Azeem et al., 2024)

Supervised Learning: Training with Labeled Data

Machine learning is a set of algorithms that learns from labeled datasets and it is a supervised learning. The model is trained on input–output pairs, where for every input data, a correct output is known. Normally in classification and regression problems, specifically, learning is done to predict outcomes over learned patterns.

Supervised learning is further divided into two subcategories:

- **Labeled Test Data**

Machine Learning Models: In cybersecurity, machine learning models are used to learn the behavior of the virus and can identify the post malicious activity. They are these models which analyze the signatures of files or analyze the structure of code or the behavior pattern of malware to detect and classify this new malware in real time.

Supervised learning is used in spam detection in email filtering systems to distinguish spam messages from normal emails. The model learns to identify common spam indicators like specific keywords, sender reputation, message formatting by analyzing labeled data consisting of examples of spam and not-spam messages.

- **Unlabeled Test Data**

Machine learning models for Anomaly Detection: unlike the traditional rule-based system, machine learning models can detect out of the box traffic to network, financial transactions or system logs. Such anomalies may be due to potential cybersecurity breaches, fraud or other irregularities.

Machine Learning in Industries: Risk Scoring is popular in many industries, for example, Finance and Health Care. For example, in credit scoring model, financial history and behavior of an applicant is evaluated to determine his or her probability of repaying a loan. Likewise, cybersecurity systems assign risk score for the potential threat, according to severity and likelihood.

Unsupervised Learning: Discovering Patterns without Labels

In the case of unsupervised learning, the model is training on data sets that are lacking pre-defined labels. The model learns hidden patterns, structures or relationships in the data as opposed to the explicit input output relationships. This is useful if data labeling is costly or not possible (i.e. for clustering, anomaly detection or data mining).

Unsupervised learning is divided into two primary techniques:

- **Clustering**

This technique is applied in different applications like customer segmentation, which is the process of grouping customers based on the purchasing behavior, preferences and demography. In the cybersecurity domain, entity classification enables the classification of entities that are possibly exhibiting similar behavior which facilitates in detecting the suspicious activities.

Clustering methods may also be used to identify strange data points that do not behave according to pre-defined clusters. As in cybersecurity, this can be utilized as a means of identifying insider threats, unauthorized access or anomalous usage patterns on a system.

- **Association Rule Learning**

This is a technique of Data Exploration i.e. in which we discover the hidden relationships between variables in the huge dataset. It is also widely used in market basket analysis, where buyers are analyzed in order to recommend related products to customers. Data exploration in cybersecurity allows security analysts to discover the correlation between many types of cyberattacks and their related vulnerabilities.

1.3.1 AI for real-time malware analysis

In defense against the more sophistication of cyber threats, it is needed to defend the critical digital assets in real time. They do not have luck with the highly dynamic malware variants like zero-day attacks and polymorphic malware. From the above it is clear that the use of Machine Learning and Deep Learning techniques with the artificial intelligence to find out the malicious patterns of network traffic along with file behavior, system log enhances real time malware analysis. The AI driven models are dynamic and continuous learning from new threats, therefore models detect and response continuously (Lewis, 2006).

The AI powered malware detection systems based upon which malware and its type are detected and classified in real time are Behavioral Analysis, Anomaly Detection and Predictive Analytics. High accuracy in classifying potential threats is achieved by both the supervised learning models trained on a large set of known malware signatures and benign file. However, it can be detected by constantly emerging malware by using unsupervised learning techniques like clustering or anomaly detection, because only deviations from normal system behavior are being made. Further strengthened, this is done through the use of Natural Language Processing (NLP) to analyze phishing emails and malicious code patterns.

AI, apart from the detection part, allows for automatic incident response and reduces the time used to mitigate an incident. With AI, you are able to have automatic alert and quarantine of infected file, and automatic deployment of patches without human help. Deep learning architectures (like CNN, RNN) are used to improve the pattern detection in malware obfuscation techniques or the like of attack strategies, via CPDs (Convolutional Pattern Decoders). Before the AI bound the real time malware analysis, companies could not get the proactively threat mitigation, following which the cyber threat of the modern cyber-attack increased.

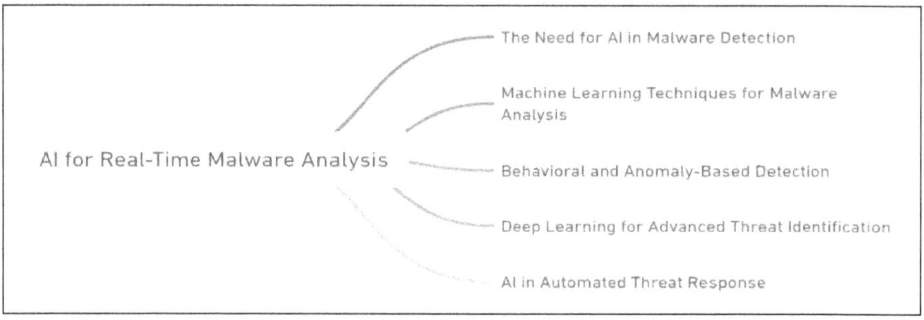

Figure 1.5 *AI for Real-Time Malware Analysis.*

Source: (self-generated)

1. The Need for AI in Malware Detection

With the increasing complexity and frequency of cyber threats, traditional malware detection techniques, such as signature-based methods and heuristic analysis, have become insufficient in identifying advanced threats. Cybercriminals are continuously evolving their attack strategies, developing polymorphic and fileless malware that can bypass conventional antivirus solutions. AI-driven malware analysis provides a proactive approach by identifying new and evolving threats in real time, reducing the risk of security breaches. AI-powered cybersecurity solutions leverage vast amounts of data, including historical attack records, network traffic, and system logs, to enhance threat detection accuracy and predict potential security risks before they cause significant damage.

2. Machine Learning Techniques for Malware Analysis

AI-powered malware detection systems leverage various machine learning techniques to improve threat identification and mitigation. Supervised learning is widely used for malware classification, where labeled datasets containing known malware and benign files help train models to distinguish between safe and malicious entities. Unsupervised learning is applied for anomaly detection, enabling the identification of previously unknown threats by detecting unusual behaviors within a system or network. Reinforcement learning further enhances AI-driven malware detection by enabling adaptive threat mitigation strategies, allowing security models to evolve and improve over time based on real-world attack scenarios. These machine learning approaches help cybersecurity professionals stay ahead of emerging threats by providing automated, data-driven security solutions.

3. Behavioral and Anomaly-Based Detection

Instead of relying solely on known malware signatures, AI-driven security solutions focus on behavioral analysis and anomaly detection to uncover malicious activities. Behavioral analysis involves monitoring real-time file activities, application behavior, and network traffic patterns to detect suspicious operations that may indicate a malware infection. Anomaly-based detection methods use statistical models and deep learning algorithms to identify deviations from normal system behavior. By analyzing system calls, process executions, and file access patterns, AI can detect previously unknown malware strains that exhibit irregular behaviors. This approach significantly enhances security defenses, providing a robust mechanism for identifying zero-day threats that would otherwise go undetected.

4. Deep Learning for Advanced Threat Identification

Deep learning architectures have significantly improved malware detection capabilities by analyzing complex data patterns at a granular level. Convolutional Neural Networks (CNNs) and Recurrent Neural Networks (RNNs) are widely used for analyzing malware executable files, network logs, and API call sequences to detect anomalies and malicious activities. CNNs are particularly effective in analyzing malware binary files by converting them into image representations, allowing AI models to classify different types of malware based on visual patterns. RNNs, on the other hand, are used for analyzing sequential data, such as time-series logs and network traffic, enabling AI systems to detect suspicious activities over time. These advanced deep learning techniques help identify obfuscated, polymorphic, and metamorphic malware, which traditional signature-based methods fail to detect.

5. AI in Automated Threat Response

AI-driven cybersecurity solutions not only focus on detecting malware but also play a crucial role in automating the threat response process. AI-powered security frameworks integrate with Security Orchestration, Automation, and Response (SOAR) platforms to facilitate real-time incident handling. When a threat is detected, AI can automatically isolate infected files, quarantine compromised endpoints, and deploy security patches without human intervention. Additionally, AI-driven threat intelligence platforms (TIPs) provide security teams with actionable insights, enabling them to take proactive measures against emerging cyber threats. The integration of AI with endpoint detection and response (EDR) and extended detection and response (XDR) solutions enhances security resilience by continuously monitoring and mitigating risks across multiple attack vectors.

1.3.2 Deep learning techniques for cyber threat identification

During the era of digital transformation, cybersecurity threats are escalating with the amount of sophistication, and the only possible way to combat such attacks is utilizing advanced and intelligent methods for detection and mitigation. Traditional rule-based systems based, e.g., on pre-defined signatures and heuristic methods are not fit to detect novel and evolving threats, especially those involved with zero-day vulnerability and polymorphic malware. For the security posture of the organization is increasingly strong, cybercriminals are developing more and more complex attack techniques, which organizations need to defend against (Gilboa & Schmeidler, 2015).

Such a rapid rise of the field of deep learning in the domain of cybersecurity is because of a very strong reason: the power of deep learning to analyze large amounts of data, pattern hunt and learn constantly, to evolve with unknown attacks and to overcome new attack tactics. With these AI-driven models, the anomalies are detected autonomously and the potential threats are predicted based on them, which becomes real time threat intelligence and consequently improves the proactive and reactive security measures. In this chapter, we analyze the keys stages pertaining to the use of deep learning based cyber threat detection system and the use of AI based system for security system keeping track of the methodologies, advantages, and actual world applications of such AI based systems.

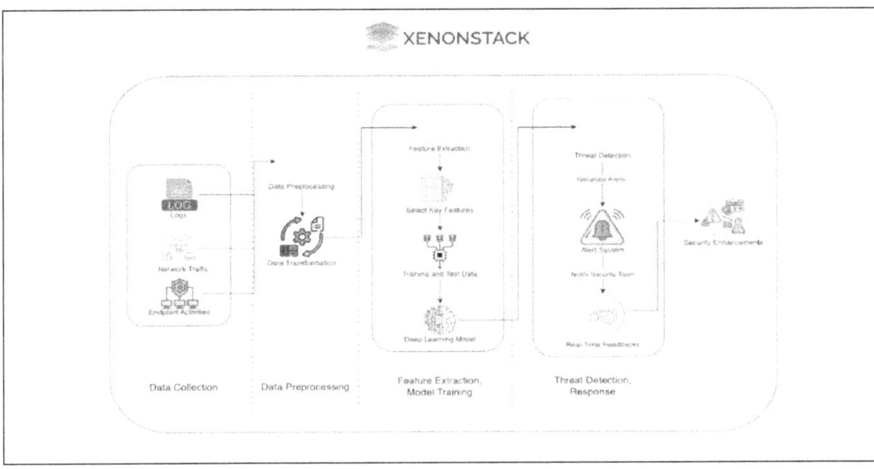

Figure 1.6 *Deep Learning-Based Cybersecurity Threat Detection Framework.*

Source: (Gill, 2024)

Data Collection: The Foundation of Threat Detection

Data collection is the first and most important step in deep learning-based cybersecurity. A powerful cyber threat detection system requires a large amount of rich data in terms of different aspects of system activities, network traffic, and endpoint behaviors. Such datasets are used as raw material for training deep learning models that will be able to differentiate between legitimate operations and malicious activities. The more diverse the data the stronger is the model that can identify threats.

Sources of Cybersecurity Data

1. **System Logs:** which record system events chronologically, and offer historical view of logins, file access history, user authentication details and security significance breaches. Using logs, deep learning models determine unauthorized access, brute force, and other strange behavior associated with an account.
2. **Network Traffic:** The first is to monitor the network packets to detect anomalies, such as data exfiltration, unauthorized communications, DDoS attacks and so on. Network data can be used to train deep learning models that identify if there is deviation in normal traffic patterns which may be intrusions.
3. **Endpoint activities:** endpoint monitoring of user interaction, software execution, file system change and process behavior. Analysis of endpoints' behavior helps in detecting malware infections, false software installations and lateral movement on a network.
4. **External threat intelligence sources:** Organizations can combine external threat intelligence sources like malware signatures, phishing indicators, known malicious IP address databases and so on to augment their cybersecurity models. By aggregating data from multiple sources, deep learning-based security systems gain a holistic view of potential cyber threats, improving their detection capabilities.

Data Preprocessing: Cleaning and Structuring Data for AI Models

Current cybersecurity data comes from the raw data that is noisy, unstructured and voluminous, which makes it very hard for deep learning models to process efficiently. Preprocessing of data is a crucial step which involves cleaning of data, structuring of data and data optimization for AI driven threat detection.

Key Data Preprocessing Techniques

1. **Cybersecurity Data:** The data is in the form of either text-based system logs, or packet captures, or numerical telemetry data. Turning these into structured datasets mean that the models of AI can be used to process and learn from them.
2. **Redundancy Removal:** As a result, large datasets often comprise redundant or irrelevant information which can deteriorate model efficiency. This helps remove unnecessary data points from the threat detection models in order to make them more accurate.
3. **Normalization of features:** As cybersecurity data usually consists of variables with different scales (e.g. bytes transferred vs. time duration), we normalize the features, so that there is no value in the model prediction due to different scales of variables.
4. **Data Augmentation:** To achieve better model robustness, greater robustness can be achieved by oversampling rare attack instances or simulating cyberattacks in controlled environment. Effective data preprocessing significantly enhances the performance of deep learning-based threat detection systems, ensuring they operate with high precision and efficiency.

Feature Extraction and Model Training: Learning from Cyber Threat Patterns

After preprocessing of the data, the next step consists of feature extraction as well as model training. The most important phase is because this is the phase that decides how much we can differentiate normal from malicious activities by a deep learning model.

Feature Selection and Extraction

1. **Key Features:** Now, security analysts select critical attributes such as unusual login attempts, unexpected network connections and unauthorized data modifications, and similar unusual user behaviors.

2. **High Dimension Equal Problems:** Slow down training or overfitting. PCA or autoencoders are utilized to efficiently reduce complexity while keeping information that is essential.
3. **Data Splitting for Training and Test:** The dataset is split into training and test sets, which helps to train the model such that it can successfully learn and, at the same time, validate against unseen threats. An 80-20 or 70-30 split is a common practice, and fine-tuning datasets are used for validation.

Training Deep Learning Models

Cyber threat detection is heavily reliant on advanced deep learning architectures.

1. **Convolutional Neural Networks (CNNs):** CNNs originally used for image recognition were used to examine network traffic and identify patterned malware.
2. **Recurrent Neural Networks (RNNs) and Long Short-Term Memory (LSTM): These models can well detect anomalies in the sequential data, for instance,** user behavior logs and time series network traffic.
3. **Transformer models:** Transformers including BERT and similar architectures make cybersecurity analytics possible by allowing to analyze the security logs at a greater abstraction level.

Learning models of security threats, anomalies, and probabilistic risk assessments are trained iteratively and increase their ability to classify security threat.

Threat Detection and Response: Real-Time Security Enhancements

Deep learning models are once trained and deployed in the real time threat detection and response systems. This means that these models are always analyzing the incoming data, find potential breaches in it and automated responses to prevent the risk.

Key Components of AI-Driven Threat Response

1. **Cyberattack identification:** The model scans live traffic on the network, logs in the systems, and user activity looking for any deviation from usual behavior which may mean that a cyberattack is taking place.
2. **Real Time Alert Generation:** If a possible threat is noticed, the system generates live alerts and categorizes the level of severity as well as suggests the right course of action.
3. **Security Systems with Automated Response Mechanisms:** The security system will autonomously respond to malicious machines trying to access the network by terminating the malicious machines' IP addresses, isolating the compromised machines to eliminate the possibility of further infiltration, or activating additional security protocols to minimize any potential threats.
4. **Continuous learning mechanisms:** They make sure that the model evolves and learns new threats over feedback loops and retraining.

Deep learning models are much more accurate and flexible than normal because they are improved by real time feedback loops to prevent cyber threats.

Advantages of Deep Learning in Cybersecurity

Some advantages of deep learning over traditional cybersecurity methods make deep learning an effective tool for fighting modern cyber threats.

1. **High Detection Accuracy:** Deep learning models have a high Detective accuracy as they can decode complex attack patterns and therefore alert at a high consultation rate instead of a high number of false positives.

2. **Scalability:** This solution scales well due to AI being able to consume vast amounts of data from enterprise level security infrastructure.
3. **Emerging Threats:** Unlike rule based static systems, deep learning learning is evolving continuously to defend against new attack strategies (zero day), since it is adaptable to the emerging threats.
4. **AI based security systems:** automated threat mitigation is an AI enabled security systems which responds to cyber-attacks in real time minimizing the damage and the manual intervention.

The deep learning integration into the cybersecurity frameworks enables the organizations to increase their efficiency in detecting, preventing, and responding to the cyber threats in a significant manner.

1.4 AI-Powered Surveillance and Biometric Security

Artificial Intelligence (AI) has transformed the operations of surveillance systems, making it possible for these devices to identify more opportunities for protection and supervision than ever before. Utilizing complicated algorithms and methods for learning patterns, security tools have become more proficient and effective (Greengard, 2021).

A closed-circuit television (CCTV) surveillance system combines equipment that involves recording, monitoring, and analyzing activities and settings for security purposes and management actions. These systems are usually used to prevent crimes – such as theft or assault – by providing evidence, maintaining public order, or enabling efficient management of premises. They are based on various technologies, such as cameras, sensors, and programs to monitor certain places or actions.

Role of AI in Surveillance and Security

AI plays a more significant role in security and surveillance, improving other security measures. Because of its transformative impact and potential capabilities, the global surveillance industry is now moving towards more sophisticated AI technology.

Among the AI applications in security and surveillance are intelligent video analytics, face recognition, asset monitoring, task automation, predictive analytics, and biometric identification. These helps detect, monitor, and respond to security threats more precisely and efficiently. Intelligent video analytics makes real-time unusual activity detection possible, reducing the need for continuous human observation. Smart face recognition improves identification accuracy within diverse environments, thereby enhancing access control and investigation processes. Asset control devices (ACDs), together with task automation, make everyday operations more seamless while maintaining round-the-clock oversight that minimizes human errors.

Surveillance Systems and Main Functions

Surveillance systems have numerous critical roles, which include:

- **Deterrence of criminal behavior:** The system aims to intimidate criminals from involving themselves in any crime. Equally, potential offenders are deterred due to the knowledge that someone will always be watching them.
- **Evidence Collection:** After a crime has occurred, the footage is crucial in law enforcement and legal proceedings, mainly as evidence. It can be used to identify suspects, understand the sequence of events, and verify alibis.
- **Public Safety:** These types of surveillance systems are found on streets, in parks, and in subway stations, all aimed at ensuring that people are protected from any kind of danger by watching over them, hence taking immediate actions at any sign of risk they may come across during their patrols across different locations.
- **Operational Efficiency:** Businesses can monitor operations through surveillance systems, ensuring that they adhere to

protocols for optimum productivity. They can, for example, monitor employee behavior, handle customer interaction, or oversee production lines.

- **Remote Monitoring:** Advanced technology has enabled modern surveillance systems to be accessed remotely, allowing individuals or companies to watch over their properties regardless of location using internet-enabled devices from anywhere globally.
- **Access Control:** In most cases, access controls are combined with building entrance and exit control systems for surveillance purposes. These systems boost protection measures by ensuring that only authorized persons can access specific parts of the premises.

Types of Surveillance Systems

Various surveillance systems perform different tasks and purposes.

Figure 1.7 *Illustrated Surveillance Camera Types.*

Source: (LINK, 2024)

These are the most prevalent ones:

- **Closed-circuit television (CCTV):** CCTV systems use analog cameras connected to a central recording device using coaxial cables. This system has been used to continuously monitor and record banks, stores, and public spaces in fixed locations. Although CCTV is known for being reliable and easy to set up, it cannot be compared to modern digital systems in terms of flexibility or advanced features.
- **IP Cameras:** An IP camera is a digital camera that sends video information over a network. Resolution is greater with IP cameras compared with regular CCTV, and image quality is also improved. Some additional functions, like facial recognition, motion detection, or access from outside locations, can also be found there. Homeowners and businesspeople have adopted these devices for various purposes.
- **Access Control Systems:** Security solutions that combine access control systems with security cameras allow for comprehensive monitoring of entrances and exits in a building. Technical equipment such as swipe cards, fingerprint readers, and smartphones, among others, achieve this.
- **Wireless Surveillance Systems:** This is the only way these systems transmit video data. The wireless technology behind these cameras allows for installing and positioning webcams anywhere. They are accommodating in areas where cabling is impossible or impracticable over Ethernet network cables. Conversely, radios tend to interfere with other devices and require high-level network availability.
- **Digital Video Recorders (DVRs) and Network Video Recorders (NVRs):** NVRs and DVRs capture footage from security cameras and keep it on tape. NVRs are designed for IP cameras, while DVRs are meant for analog CCTV. Each recorder model has different storage options, playback, and remote-control alternatives.

- **Cloud-based Surveillance:** Systems located in the cloud store video material in remote servers, meaning data can be easily accessed, shared, and managed. This eliminates any form of storage device at any location, saving hardware costs while facilitating more security due to encryption mechanisms that are used both during transit periods and when kept at rest.

Integration of AI with Surveillance Systems

By integrating AI with surveillance systems, the security is increased with threat detection in real time, facial recognition, if any anomaly is there somewhere. AI powered cameras can analyze large amount of video footage trying to identify suspicious activities and lowering the requirements of constant human monitoring. In many industries, this advanced technology has increased response times, decreased false alarms and overall efficiency of surveillance.

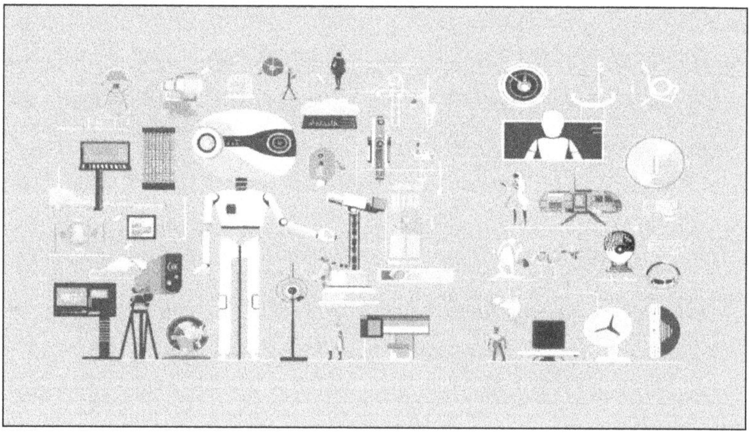

Figure 1.8 *AI and Robotics in Surveillance and Security.*

Source: (LINK, 2024)

The integration of AI and surveillance is mostly done in the following ways:

1. **Cameras and Sensors:** Cameras have AI algorithms and sensors within them to improve data multitasking. These smart cameras can focus by themselves, follow things up, and optimize the quality of images based on the kind of scene they detect.
2. **Video Management Software (VMS):** An AI-based video management system (VMS) verifies patterns in videos, both live and recorded, to sound an alert on potentially troubling activities.
3. **Cloud-Based Solutions:** With cloud-based solutions, one can implement the capacity of AI in their web apps, make use of scalable computing and storage capacities, and optimize big data processing, which is a great deal more efficient. This would mean complex analysis-based manipulation of information as well as integration across various databases.
4. **Edge Computing:** Artificial intelligence algorithms are carried out at the edge, that is, on the device itself or very close to where the data originates. As a result, less delay is experienced, and less bandwidth is utilized, considering that information is assessed nearby.

1.4.1 AI-driven facial recognition and behavioral analysis

Facial recognition and behavioral analysis has merged into artificial intelligence (AI) in augmented surveillance systems based on intelligence that's much more effective in security, law enforcement, and public safety. Facial recognition driven by AI is an entire process of scanning and mapping out facial features with the help of deep learning algorithms for speedy identification of individuals. From crime prevention to suspect tracking and all the ways from there, this technology is used everywhere to provide unmatched accuracy for face recognition even under adverse conditions like poor lighting or partial obstructions. AI powered systems allow using neural networks and a computer vision to keep improving accuracy as well of work with changes in facial expressions, aging or disguises.

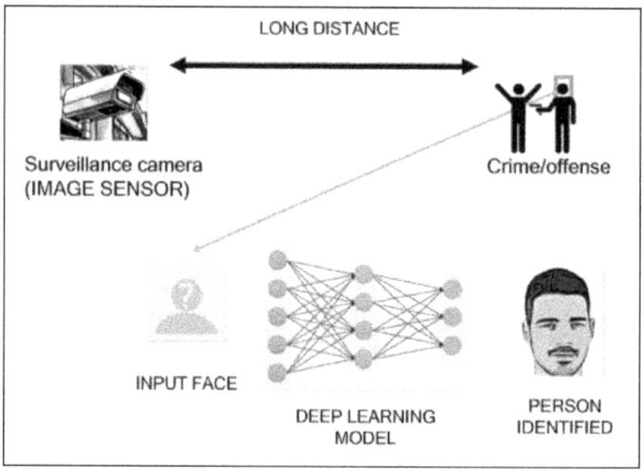

Figure 1.9 *AI-Powered Facial Recognition in Surveillance.*

Source: (Llauradó et al., 2023)

Beyond facial recognition, AI behavioral analysis goes further once you do to identify movements, gestures, and pattern of humans, and pick up the anomalies or suspicious activities. Real time crowd behavior can be assessed, movement can be identified as erratic and can be predicted as to what security threats exist. For example, an AI system gauging a public space can recognize such factors as aggressive behavior as well as unattended objects or a pattern of loitering that could signal a crime. The security personnel now have these insights and can intervene proactively to reduce risks before the incidents escalate (Ferdin Joe John Joseph, Karthikeyan Chinnusamy, P. Paramasivan, R. Regin, 2024).

Face recognition, and subsequently, the integration of AI into this practice, involves ethics and privacy issues, which need to be regulated strictly as well as having transparent policies. Data protection, information about preventing biases in AI models and the implementation of surveillance measures based on consent are crucial for good deployment. With the development of technology, it is critical that there is a proper balance of security and individual rights to develop trust and the ethical usage of AI driven surveillance systems.

1.4.2 AI in video surveillance and anomaly detection

Video surveillance with AI integration has transformed the traditional security monitoring into a proactive and therefore intelligent and in time detection of threats. Deep learning and computer vision are employed by deep learning to support varieties of AI driven surveillance systems to analyze live video feeds, spot potential risks, and also sense unusual routines of behaviour. Unlike conventional surveillance methods that depend so much on human monitoring, AI could reduce the time it takes a security personnel to respond to suspicious activities and make them less work by automatically flagging them (Yan, 2019).

Figure 1.10 *Surveillance Footage of Criminal Activities.*

Source: (N.A. et al., 2025)

AI powered surveillance is not possible without anomaly detection for the system to detect deviations from normal behavior patterns. AI can identify such anomalies as an unauthorized access to a system, sudden movement, loitering, or aggressive behavior if it analyzes a large dataset of movements in various environments. This is very useful in high security areas, public spaces and industrial areas where quick intervention is important. Moreover, AI can be connected to facial recognition and object detection for more security, for instance, to detect known threats or missing individuals.

In addition to security, AI is applied to video surveillance to manage traffic, retail analytics and safety at work places. AI driven cameras in smart cities monitor traffic flow, detect accidents and optimum traffic signal that make the movement smoother. These systems are useful in retail activities where they assist in analyzing customer behavior to improve customer service and prevent theft. With the rise of AI technology, video surveillance and anomaly detection will become more and more important in protecting safety, security and operational efficiency in the different industries of the world.

1.5 Ethical and Privacy Concerns in AI Security

The integration of rapid AI into security systems has been cause for a lot of ethical and privacy concern. Also, the usage of AI based surveillance, facial recognition and threat detection technologies to make the safety improved has its risk of mass surveillance and misuse. It could be argued that arguably the biggest dilemma is that of security against the movement of personal information. AI powered monitoring systems can also collect a vast variety of personal data about an individual without the explicit consent of the person. Despite having such systems, governments, corporations or malicious actors could be allowed to use personal information and violate civil liberties and fundamental rights (Omar Santos, 2024).

A third big problem is bias and fairness in AI security. Using historical datasets that may potentially include racial, gender or socioeconomic wealth biases, an AI model is trained. In fact, these can be discriminatory, which means that they might lead to a wrongful identification through facial recognition or an unfair risk assessment in predictive policing. The studies have also proven that the AI security tools usually misidentify people trained from a minority group more than they do others, which could become legal and also social wrongdoings. To ensure that the AI security systems stand for fairness and safety, the inequalities must be clearly defined and there must be clear means of transparency for auditing.

To arrive at this, there is a need for an ethical framework for AI security that focuses on transparency, accountability and the user right as priority. Another one of the simplest ways to do that is by applying privacy preserving techniques like differential privacy and federated learning. It is also required that the GDPR and AI ethics guidelines should be enforced for the deployment of responsible AI. Such policies should consider public awareness and stakeholder engagement for the promotion of policies that are both security and ethical oriented. AI also has a responsibility to protect from threats, without compromising on privacy, and to ensure that technology is being conducted appropriately according to ethical and legal standards.

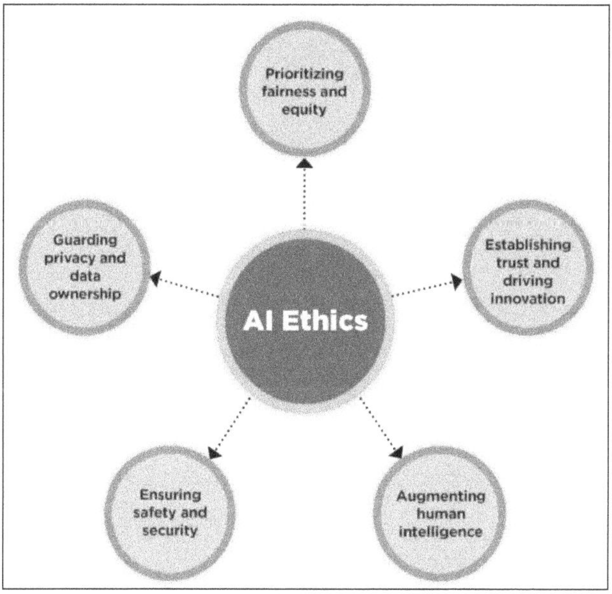

Figure 1.11 *AI Ethics and Its Key Principles.*

Source: (Intersystems, 2023)

The image is a conceptual framework of AI Ethics, which consists of five principles that are necessary for the responsible and fair development of artificial intelligence systems. They are interrelated building blocks to guarantee that AI technologies are deployed in a secure way that will promote the interest of society, but that will address the associated risks.

1. **Ensuring AI Fairness and Equity** – AI systems are built for fairness and equity of treatment among the diverse user groups. It entails using data across different datasets, clear algorithms, fairness audits to avoid discrimination of race, gender or socio-economic status.
2. **Driving Innovation and Creating Trust** – As is the case in any area of technology, ethical AI should be transparent and explainable to its users in order to establish trust. The adoption and innovation with AI can be ensured by stakeholders by making sure that AI models are understandable and accountable.
3. **Augmentation of Human Intelligence** – It should be used to improve human intelligence rather than replace it. The goal of ethical AI development is to empower individuals and professionals in all industries with AI insights and ensure human oversight in severe decision-making processes.
4. **Safety and Security** – The system should be immensely safe, secure and be protected from cyber threats. Safety is the main focus of ethical AI and the development of safety measures that prevent such harmful consequences as adversarial attacks, data breaches, and unauthorized surveillance.
5. **Data Protection and User Privacy** – Stressing on the privacy of the users and data ownership. Strenuous data security measures must be adopted by the organizations to make sure people have control of their personal data and AI cannot be used in mass surveillance or data exploitation.

By going through these principles, AI can evolve and be deployed in a manner that fits the human values, hence, minimizing the ethical risks while the innovation is engaged in a responsible manner.

1.5.1 Balancing AI security with data privacy

The merging of AI in the critical systems continues and therefore, it is imperative to find balance in AI security and data privacy. The amounts of data that occur in practice for AI systems to work effectively are immense in cases such as predictive analytics, cybersecurity, and personalized services. But the large amounts of sensitive user data collected, stored and processed are risky in terms of privacy, including unauthorized access, data breaches and misuse. To ensure that AI is safe and people's privacy is protected, we need to use strong encryption, anonymization processes and strictly control activities with personal information to prevent it from being revealed to who should not see (Puplampu, 2024).

The challenge of balance in this one is to keep AI models accurate and effective while following the privacy laws. Differential privacy, secure multi party computation, and federated learning are becoming a suite of methods to train the AI models using the decentralized datasets in a way that bypasses the access to or storage of such information. The regulatory frameworks such as GDPR (General Data Protection Regulation) and CCPA (California Consumer Privacy Act) mandate what the organizations need to be transparent about the data collection and processing and how users would have a control over their data. They force AI developers to design the system so as to achieve the highest possible security and privacy while keeping the performance at the highest levels.

Ethical AI development also contributes significantly to dealing with issues of privacy. Fairness of AI driven security mechanism such as anomaly detection or fraud prevention must be ensured avoiding the biases that allow discrimination or misuse of personal data.

This means, organizations must have a privacy by design approach where they integrate security into AI architectures rather than as an afterthought. The right balance is to have public trust in AI, which means that security solutions powered by AI are effective while respecting the fundamental right to privacy. By adopting transparent policies, the best use of advanced privacy preserving technologies, and the effective management of AI security and data privacy, the chasm between data privacy and AI security can be managed well.

1.5.2 Regulatory challenges in AI-powered security solutions

Artificial intelligence is integrating into security solution in order to bring down the ways organizations prevent and respond threats. The security systems with the help of AI, such as facial recognition, biometric authentication, predictive analytics, autonomous threat detection have greatly improved surveillance and have real time monitoring and proactively, defending. However many of these technologies have been widely adopted, and they present a suite of problems relating to the regulatory, ethical and legal with which they must also come to grips with to be used in a responsible and compliant manner. While governments, policymakers and industry leaders are trying to find the balance between technological advancement and privacy protection, civil liberties, corporate accountability and so on (Bojan Kolosnjaji, Huang Xiao, Peng Xu, 2024).

With more and more sophisticated AI security solutions, they need access to vast amounts of personal and behavioral data to work effectively. The result of this is the violation of data privacy, surveillance ethics, algorithmic bias, and cross border compliance regulation. Worse still, because the absence of univocal global AI regulations make the situation even more complicated, different locations impose different laws and restrictions to use AI based security implementations. To achieve what is still imperfect yet necessary defense for the world, organizations must steer through these regulatory landslides, ensuring that AI-driven security technologies are robust and secure, but at the same time in compliance with the law and the best ethical practice.

In the following sections, the key regulatory challenges that organizations, governments as well as AI developers face when deploying AI based security solutions are explored. Such issues include data privacy compliance, ethical issues with AI based surveillance, cross border conflicts of regulations, legal liability and accountability and the balance between security and civil liberties. By knowing about these challenges, the stakeholders can make efforts to design responsible, clear, and legally sound AI security framework that will not compromise safety and will respect fundamental rights.

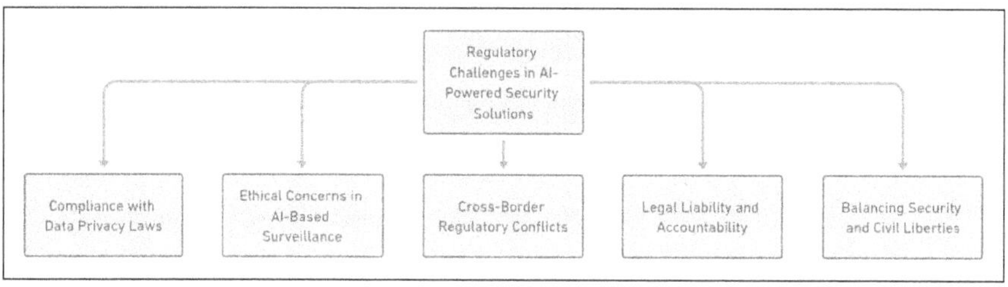

Figure 1.12 *Regulatory Challenges in AI Security.*

Source: (Self-generated)

1. Compliance with Data Privacy Laws

Many AI-driven security systems are prone to collecting, analyzing and storing personally identifiable information (PII) that makes them subject to stringent data protection regulations like the General Data Protection Regulation (GDPR), California Consumer Privacy Act (CCPA), Personal Data Protection Act (PDPA) etc. When AI models depend on continuously access to individuals' sensitive data for training and optimization while abiding by the rights of individuals to privacy and consent to obtain systematic compliance are challenging. In order to meet regulatory standards, organizations deploying AI powered security solutions must be using robust data governance policies, encryption methods, and an anonymization technique (s) in play to employ strict methodologies to carry out such deployments.

2. Ethical Concerns in AI-Based Surveillance

Ethical dilemmas arise as AI is being more and more applied to public and private surveillance, specifically with regard to consent and bias. For instance, facial recognition has been accused for misidentification and racial bias, the fear of unfair surveillance and wrongful arrest. AI surveillance tools must be regulated so that the rules are clear and contentious outcomes are avoided, including discriminatory outcomes, and in the way that it does not infringe upon civil liberties. To minimize ethical risks of the security applications based on the AI it is required to monitor it with transparency audit mechanisms and human oversight.

3. Cross-Border Regulatory Conflicts

There are different AI security regulations across the different countries and this poses compliance challenges for global organizations. Different from China, European Union, and the United States, international standardization makes security leveraging of an AI tends to be complicated through cross-border implementation of AI powered security solutions. If a company operates in different jurisdictions, it has to deal with regulating requirements that may give rise to legal risk and result in operational inefficiencies. To foster innovation with consistent security and privacy protections, but to do it harmonized and on a global scale, there has to be development of harmonized global AI regulations.

4. Legal Liability and Accountability

Determining legal responsibility in regard to AI driven decisions or failures is one of the most pressing regulatory challenges in the area of AI based security. If an AI based security system identifies a suspect incorrectly, does not detect a security breach, or goes beyond its purpose, a natural question is asked, who is responsible, the AI designer, the organization deploying the system or the end user. The liability structures in the regulatory framework should be clearly defined, while the explainable AI (XAI) and human in the loop approach should be used to make the systems accountable in security operations.

5. Balancing Security and Civil Liberties

Security solutions based on the power of AI has to balance the well-being of the public and their rights. The excessive use of such security measures as mass surveillance or predictive policing may cause the violation of the human rights and the privacy laws. Such regulatory frameworks would need to insure that AI security technologies are, themselves, not used by governments, for unlawful surveillance, discrimination or political oppression and would not enable this to happen. And guardrails that establish both can help create ethical AI policies and independent review bodies to keep AI from oversaturating security and still allow it to make security contributions.

Multiple Choice Questions (MCQs)

1. What is a primary benefit of AI in security?

A. Eliminates all cyber threats

B. Enhances real-time threat detection and response

C. Replaces the need for human cybersecurity experts

D. Reduces the cost of all security solutions

2. How does AI contribute to intrusion detection and prevention?

A. By manually analyzing logs

B. By automating anomaly detection and threat response

C. By eliminating the need for firewalls

D. By blocking all network traffic

3. What is AI-powered threat intelligence primarily used for?

A. Encrypting network traffic

B. Predicting and mitigating potential cyber threats

C. Automating software updates

D. Replacing traditional antivirus programs

4. Which AI technique is most effective for real-time malware analysis?

A. Decision Trees

B. Deep Learning

C. Manual Signature-Based Detection

D. Regression Analysis

5. How does deep learning enhance cyber threat identification?

A. By creating static security policies

B. By detecting patterns and anomalies in vast datasets

C. By manually reviewing threat reports

D. By blocking all traffic from unknown sources

6. Which technology is commonly used in AI-powered biometric security?

A. CAPTCHA verification

B. Facial recognition and behavioral analysis

C. Firewall-based security

D. Encrypted SMS authentication

7. What is the role of AI in video surveillance?

A. To completely replace human security personnel

B. To identify anomalies and suspicious activities in real time

C. To store unlimited amounts of video footage

D. To disable surveillance systems when not needed

8. Why is balancing AI security with data privacy important?

A. To ensure AI systems do not overconsume computing resources

B. To comply with regulations and protect user rights

C. To limit AI's ability to detect threats

D. To prevent AI systems from evolving

9. Which is a key regulatory challenge in AI-powered security solutions?

A. Lack of computing power

B. Cross-border regulatory conflicts

C. AI's inability to detect patterns

D. Manual threat detection becoming obsolete

10. What is a common ethical concern in AI-powered security?

A. AI replacing all security professionals

B. AI making security measures too expensive

C. AI being used for mass surveillance and violating privacy rights

D. AI making cybersecurity laws unnecessary

Answers

1	2	3	4	5	6	7	8	9	10
B	B	B	B	B	B	B	B	B	C

Chapter 02

AI-Enhanced Enterprise Resource Planning (ERP)

2.1 Introduction to AI in ERP

Artificial intelligence (AI) on the other hand is the integration of artificial intelligence (AI), which includes machine learning (ML), natural language processing, predictive analytics, into an enterprise resource planning (ERP) system. These AI tools can help people decide better, automate the repetitive tasks, and perform invaluable sophisticated analysis of data along with data forecasting. The goal of ERP with AI is to make operation more effective and improve business processes. Using AI, ERP systems can analyse massive amounts of structured and unstructured data at a moment's notice, providing companies with a better insight on the operation. AI driven automation helps by automating processes and thereby reduces manual labor, minimizes human error, but also increases output. Also, intelligent chatbots, tailored suggestions, automated help systems and so on, are other ways to help customer care through AI powered ERP. Artificial intelligence (AI) in supply chain management helps in forecasting changes in the demand with the help of market trends and meteorological conditions and in optimizing the inventory levels and reducing delays. Financial reporting, irregularities detection, and stop of fraud may be automated in accounting and finance through AI. AI based ERP solutions can be used in workforce planning, employee performance analysis and talent acquisition in the human resource management. As AI is developing, ERP systems become smarter and more flexible and provide businesses with the opportunity to make data driven decisions, save money, and get competitive advantage. In essence, AI in ERP tries to make enterprises more flexible, responsive and future ready, individual tasks are no longer the only thing to be automated. (Moallem, 2021).

AI Work With an ERP

Integration is where it all begins. The ERP system and AI algorithms are combined to provide intelligent automation. Employees will be able to focus on other things since they will have more time for strategic and valuable activities.

Furthermore, AI systems may learn from past data and forecast or suggest various outcomes, such as supply chain, inventory, and customer expectations. Over time, this improves your operational efficiency and helps you make decisions faster.

The items that follow provide a clearer explanation of how AI functions with ERP:

- AI algorithms are fed a lot of data about the business, for instance, sales numbers, customer buying habits and the market trends. It allows businesses to see what patterns, what insight can help them to take decisions.
- Being able to represent in natural form, AI powered chatbots can better comprehend and reply to the consumer inquiry, this means real time support and help.
- ERP systems are made to learn and improve by machine learning algorithms, constantly learning and adapting to user's interactions and inputs in order to get more accurate and efficient over time.
- AI algorithms can replace employees in conducting routine tasks like data entry, inventory management, invoice processing, thereby reducing human errors time and saving time.

- Finally, AI enabled predictive analytics enables in predicting demand and reducing inventory levels, as well as improve supply chain management.

In conclusion, utilizing the potential of AI in ERP systems allows organizations to increase productivity, accuracy and agility, and that consequently helps to make better decisions and this positively impacts business outcomes.

Advantages of Implementing AI in ERP Systems

Given above, artificial intelligence in ERP systems has changed the business world, at least in terms of the outputs of ERP systems.

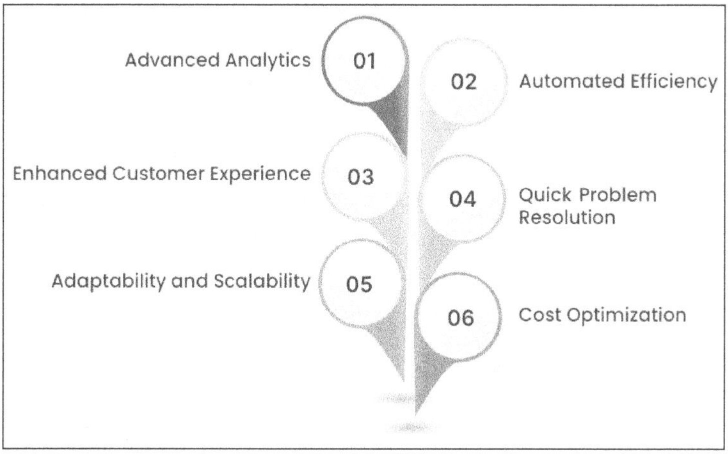

Figure 2.1 *Advantages of Implementing AI in ERP Systems.*

Source: (Paras Kela, 2024)

1. **Advanced Analytics**: Rich analytics is the first significant advantage of AI in ERP. With the use of AI software, large volumes of data may be precisely and swiftly analysed by modern ERP systems. It boosts corporate intelligence, forecasts, and identifies patterns, allowing for the establishment of strong data-driven business decisions. All things considered, AI in ERP is being used to stay one step ahead of the competition.

2. **Automated Efficiency**: One of the next benefits of AI in ERP is that they help automate mundane tasks. AI based ERP automates repetitive such activities like data entry, report generation, inventory management, and so on. It sets different priorities and frees up a great deal of time and resources that you can use to expand your business through strategic initiatives.
3. **Enhanced Customer Experience**: ERP systems with AI offer an overall view of your customers. It is a consolidation of data from different points allowing you to personalize the experience. You can provide your customers better service by having tailored product recommendations, timely support, and so on, which will result in long lasting relationships.
4. **Quick Problem Resolution**: AI is a master problem-solver. AI in ERP systems, using machine learning algorithms, identifies the unexpected pattern, detects the inefficient processes, and notify you about them in real time. It even recommends the best ways to handle problems without disrupting your business operations and saves you a good deal of time.
5. **Adaptability and Scalability**: ERP systems with AI are able to learn and adapt easily. The use of AI algorithms means as your business develops, those algorithms continue to optimize processes, adjust recommendations, and also to fine tune accuracy to improve it. This means that an AI based ERP system can grow along with your business no matter whether you are expanding into new markets or revising your customer support approaches.
6. **Cost Optimization**: Finally, AI in ERP software solutions aids in optimizing costs throughout your organization. It improves inventory management and reduces wastage by analyzing historical data and sales pattern. Eventually, these cost saving measures also help in achieving higher profitability as well as thereby improving the financial performance.

In sum, AI and ERP together are a powerful synergy that provides your business from the top down.

2.2 AI-Driven Automation of Workflows and Decision-Making

Today in this fast world, corporations should always be working on their workflows and the best practices of decision making so that they can live in this rat race. The companies who will be winners of the 'AI Change Agent' award have embraced AI-driven automation in their Enterprise Resource Planning (ERP) and have connected it with intelligent technologies — robotic process automation, natural language processing, and machine learning.The advent of these advancements has empowered ERP systems to simplify mundane jobs, decrease the probability of human mistakes, as well as speed up taking decisions from the insights gathered from data (Coplin, 2024).

ERP with AI automation saves operational time that automated processes remove from repetitive tasks like filling up data, making invoices, and inventory management. This can be used rather than depending on manual efforts and this AI algorithms can intelligently analysis data discover patterns, and act accordingly to predefined actions with least oversighting. This not only helps the organizations saving resources and time, while also enabling the workforce to concentrate on value added work and not on petty administrative work.

AI Workflow

An AI workflow is a series of processes that use artificial intelligence (AI) to complete a task. This could involve anything from simple data entry and analysis to more complex image recognition. AI workflows aim to automate repetitive, time-consuming, or error-prone jobs, increasing efficiency and productivity and allowing you to focus on more strategic and creative work. They are crucial in streamlining and future-proofing your business operations.

AI Enhance Traditional Workflow Automation

Intelligence and adaptability is what AI brings to the picture when compared to sheer automation on the workflow. Conventional automation can perform as per predefined rules but AI powered workflow can grub data in bulk, learn from patterns, make necessary intelligence decisions at real time.

AI workflows take advantage of the technology of machine learning, neural networks and others that enable the optimization and adaptation to changing conditions and continuous improvement of performance. This can be done using AI driven automation and you will be able to do sentiment analysis, predict possible bottlenecks, suggest solutions, do nuanced decisions, etc. It encourages innovation and sustainable growth of business.

Benefits of AI Workflows for Businesses

AI workflows enable you to develop better organization's data management strategies and act as value added solution in long run. Some important things about using AI workflows are as follows:

Figure 2.2 *Benefits of AI Workflows automation.*

Source: (airbyte., 2024)

1. **Increased Efficiency and Productivity**: AI workflow automation will help you to get the faster turnaround time

and better resource allocation. This helps speed up project timelines as well as let your team's pay attention to higher order initiatives that allow for problem solving, overall productivity.

2. **Cost Cutting**: AI workflow management takes the place of manual labor and reduces the cost of operation by cutting operational costs directly. It decreases the probability of errors that will result in expensive rework or delay. In addition, long term savings are achieved from AI-driven optimization that helps identify gaps in data processes.
3. **Improved Accuracy and Quality**: AI workflows enable organizations to handle the huge amounts of data while simultaneously adhering to high standards of quality and precision. This helps to minimize the discrepancies and to comply better with industry standards and regulations.
4. **Scalability and Flexibility**: AI workflows can very easily scale either up or down as the workloads vary without compromising efficacy, speed or data quality. Their flexibility depicts that they are flexible enough to transact quickly with changes in market conditions, regulatory policies, etc.
5. **Enhanced Decision-Making**: AI analysis helps organizations to find trends and patterns, allowing them to make the right decision based on actual and timely information. By doing this, they can be proactive in finding the opportunities, mitigating risks and taking the right decision to align to long term strategic goals.

2.2.1 Intelligent process automation (IPA) in ERP

Intelligent Process Automation (IPA) is a revolution in how the Business Process is being done through artificial intelligence, machine learning and robotic process automation to accelerate chores in the Enterprise Resource Planning (ERP). Compared to its traditional counterpart, IPA allows the ERP systems to work in a dynamic manner, and respond to the data patterns and take intelligent decisions.

In effect, this evolution is allowing businesses to improve efficiency, cut down operational costs and increase productivity across the board (Matthews & Greenspan, 2020).

The objective of IPA in ERP is to automate repetitive, laborious tasks like data entry, invoice processing, procurement approvals and payroll management. With the help of AI driven bots, these tasks can be done more accurately and faster without let the humans make any errors and spending their time to work on some higher value work. For instance, an AI enabled ERP system can extract data from invoices, validate information with purchase orders, process payments without any manual intervention and so on. Moreover, this not only speeds up the workflow but also helps to comply with the financial regulations.

Through providing intelligent insights about Making decisions in real time using historical data and predictive analytics, IPA also comes with an additional value. For example, in supply chain management, such a system where ERP database is driven by AI can keep track of inventory level; predict any demand fluctuations and then automatically restock, re-distribute the goods as needed. Similarly, IPA can be used as a human resource application in the process of hiring, screening resumes, scheduling interviews etc. in a smooth way.

By applying the IPA in ERP, the businesses achieve the advantage of streamlining the operations, enhancing the accuracy, and increasing the agility. IPA will only become more important as it plays an even larger role in the future of enterprise resource management in shaping the smart, fast, and efficient way in which processes are made.

Benefits of Intelligent Process Automation (IPA) in ER

Integrating IPA into an ERP system allows for a great number of advantages: it helps to speed up ERA while also guaranteeing its level of accuracy and the decision making within it which, as a result is revolutionizing the way of working of a business. The following are the main advantages of an IPA in ERP:

1. **Reduced Time and Effort:** IPA automates repetitive rule-based tasks like data entry, invoice processing, as well as report generation leading to highly increased efficiency and productivity. By doing this, employees can concentrate on the strategic, high value activities that contribute to business growth.
2. **Reductions in Costs**: Businesses can reduce their operational costs when they automate the processes, such as the use of manual labor, as errors and delays. For instance, in the case of IPA, it reduces the requirements for complex back office, which allows for cost savings in administrative, compliance management and customer support tasks.
3. **Human errors in data processing**: calculations, and documentation results in errors, which in turn leads to the costly mistakes. Handling, validating the data, reducing discrepancy errors and being compliant to regulatory standards is made easy by IPA.
4. **Reduced IT Cost with Big SQL**: Big SQL technology has reduced the cost of maintaining the server, cutting the time and money spent on managing servers with traditional methods down to a minimum. Thc application of ΛI driven automation in, be it supply chain management where the demand forecasts need to be predicted, or for contract management related financial planning, eases out these smarter business strategies.
5. **Improved Compliance and Risk Management**: Regulatory compliance is a crucial part of running a business and IPA does this by automating documentations, audits and reports. Automation on their behalf can detect anomalies, alert about the possibility of a risk, and confirm the fulfilment of legal and financial regulations.
6. **AI Powered Chatbot:** Automated Order Processing and Personalized Recommendations: IPA helps businesses provide

better customer service by adding AI into the form of chatbots, product order processing and personalized recommendation. In light of higher customer satisfaction and loyalty, faster response time and seamless service delivery would be provided.

7. **Scalability and Flexibility**: With growth, manual processes are more and more difficult to manage. By connecting with ERP systems, IPA allows them to scale naturally from increased data volumes and operational complexity without extra human resources.
8. **Automating the Routine Administrative Tasks**: IPA automates the routine administrative tasks freeing the human resources and thus they can engage in innovation, problem solving and decision making that improves overall business performance.
9. **Ease of Integration with Existing Systems**: IPA tools can integrate with the existing ERP systems, which lets the companies take the advantage of automation without having to break the backbone of the entire infrastructure. It saves you from unnecessary expenses on making your old technology to work with AI.

2.2.2 AI-powered decision support in business operations

As a matter of fact, Artificial Intelligence (AI) is changing processes in business operations through improved decisions support systems, making business operations more efficient and accurate without sacrificing strategic planning. Historical data and human intuition-based decision-making process might not always be the best means of arriving at a decision. The AI enabled decision support system (DSS) are machine leaning, prediction analytics with real time data processing that integrate with the businesses to support making decisions faster and more effectively (Vemuri, 2020).

Extracting data from large number of sources is another key benefit of AI in the decision-making process as AI is capable of analyzing vast amount of structured and unstructured data and looking for trends, patterns and correlations over multiple sources, which may not have been visually noticed by man. This makes it possible for the business to forecast market changes, optimize its resource management and mitigate its risks in a proactive manner. For instance, in financial planning, an application of AI for DSS can figure out various cash flow trends, detect strange transaction, highlight budgetary alterations for getting the optimum profitability. Similarly, AI can also forecast demand volatility in supply chain management, suggest the most suitable inventory levels, as well as identify impending disruptions before they affect the operations.

Along with that, customer relationship management (CRM) has been transformed by AI by realizing the behavior, preferences, and feedback of the customer in real time. This lets businesses personalize intended marketing campaigns, boost the involvement of customers as well as retain their clients. In addition, AI driven automation in HR allows an organization to make data backed hiring decisions, assess employee performance and predict workforce trends.

Combining AI into the decision-making process allows businesses to operate with more acceptable operational efficiency, increased agility and competitive advantage in factoring increasingly data production. Besides helping executives to make smarter strategic decisions, AI allows employees to be guided by intelligent recommendations as the culture of innovation and continuous improvement.

Artificial Intelligence (AI) is taking the decision-making process of businesses to next level by enhancing data driven processes. It filters and examines massive quantity of datasets in real time to discern patterns, trends and correlation that matters in strategic decision making. This diminishes the need of intuition and human processing for making decision, and thusly fixes it to be much more dependable and objective. AI powered with predictive analytics adds more strength to it, giving forecasts for market trends, customer needs and financial risks on the base of historical and real time data. This allows businesses to take a proactive stance against challenges and seize on business opportunities in advance, putting themselves on a competitive footing.

Besides strategic decision making, AI enhances the operational efficiency by making better resource allocation and supply chain management. This allows for very calculated budgets, workforce planning, and minimum waste for the maximum return on investment. ERP with AI automatically analyzes cash flow and offers budget plans and revenue growth forecast. AI in the supply chain forecasts the demand fluctuations and keeps the right inventory levels, and it also automates procurement by applying real-time supplier performance and the market trend. Additionally, AI based tools in CRM helps in measuring customer behavior and feedback to suggest best personalized services, suggest best marketing strategies and increase customer loyalty.

Both routine decision making and support functions in an organization depend on AI. By analyzing transactional patterns, it identifies the fraud and possible risks and thus helps in reducing the financial losses and ensuring both regulatory and internal compliance. In HR, AI helps in finding talents, analyzing employee's performance, and also predictions of turnover where it provides data backed retention strategies. Powered by AI, real time dashboards help executives get continuous insights on important performance indicators (KPIs), thereby enabling the response to events occurring in real time in an accurate manner. AI enables you to automate repetitive decisions that square away price, restock and workflow approvals so that human talent can stay fully utilized doing more strategic worth-driven tasks, which charges the productivity and innovation.

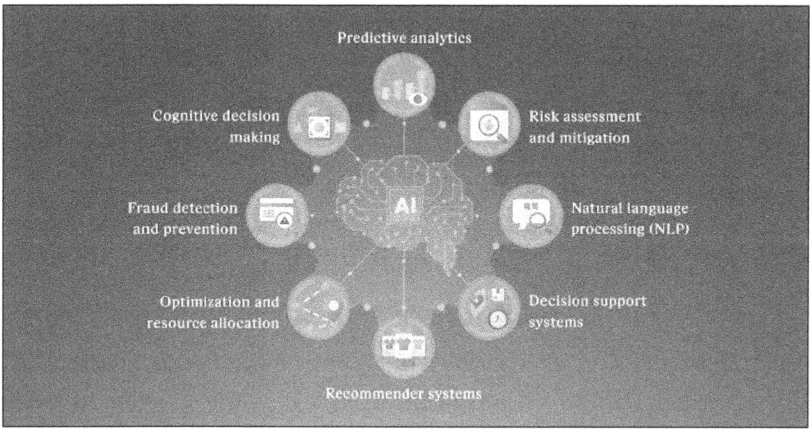

Figure 2.3 *Applications of Artificial Intelligence in Decision-Making and Analytics.*

Source: (Ambassadors, 2024)

1. Predictive Analytics

Predictive analytics leverages historical data, statistical algorithms, and machine learning techniques to forecast future events and trends. By analyzing patterns in large datasets, AI systems can anticipate outcomes such as customer behavior, equipment failures, sales forecasts, or market dynamics. This empowers businesses and institutions to make proactive, data-driven decisions, reduce uncertainty, and optimize planning processes. Industries such as finance, healthcare, and retail increasingly rely on predictive analytics to stay competitive and agile in a rapidly changing environment.

2. Risk Assessment and Mitigation

Risk assessment tools based on artificial intelligence examine a large number of criteria and detect possible threats and risks in operations, finance, or cybersecurity. With real time data and sophisticated models, these systems can assess the credit worthiness, detect security breach or evaluate the risks on the compliance measures. Apart from recognizing the risk, AI can suggest ways to try and prevent/limit that risk by simulating scenarios and show how much it costs to implement preventive actions. Most however, are particularly useful in fields like banking, insurance, supply chain management, infrastructure management etc. where risk management is vital for the sustainability and growth of the business.

3. Natural Language Processing (NLP)

AI refers to a group of subfields which acts up to simulate intelligence, one of them being 'Natural Language Processing (NLP)', which refers to computers 'understanding, interpreting, and generating human language in a meaningful way'. NLP powers applications such as chatbots, virtual assistants, translation services, and text summarization tools. It enables the organizations to automate customer support and perform sentiment analysis as well as extract insights from the unstructured text data such as emails, reviews, and social media posts. NLP fills the communication between the machine and human and make AI more approachable and user friendly for day-to-day applications.

4. Decision Support Systems

The decision support systems (DSS) enhanced by AI allow people as well as the organization to make data informed decisions which analyze the large volumes of data, find patterns, and gives actionable insights. In healthcare they are used to support such things as clinical decision making, in business to support strategic planning and operations. AI powered DSS differs from the conventional tools since they can be updated to changing data sets, learn from the past decisions and get constant improvements, thus boosting the accuracy and the confidence in the decision.

5. Recommender Systems

AI algorithms are used in order for recommender systems to personalize the content, the products and the services that are presented to the user depending on the user's preferences and the user's behaviors and interactions. Used widely in e commerce platform to provide best possible items on suggestions, used widely in entertainment service to curate content, used in education to tailor learning path items etc. Relevant suggestions that these systems deliver are based on collaborative filtering, content-based filtering or hybrid techniques thus improving user engagement, customer satisfaction and revenue generation.

6. Optimization and Resource Allocation

AI is important for enabling an organization to optimize and allocate resources as much as possible. It is a highly dynamic system that can dynamically distribute personnel, schedule production runs, manage supply chains in real time, and do the same for energy or materials. Considerably different from human, AI models systematically process a numerous of variables and constraints at once, quickly and accurately foretelling what is most effective. As this is particularly beneficial in industries like logistics, manufacturing, and production, this is utilized for the optimization of resources.

7. Fraud Detection and Prevention

The aim of the AI based fraud detection systems is to detect and prevent all types of deceptive or illegal activities through real time monitoring of transactions and behavioural patterns. They are anomaly detection, pattern recognition, and predictive model type systems which use to flagning suspicious activities from normal behavior. Such tools are used by financial institutions, online retailers and cybersecurity firms to prevent credit card fraud, insurance scams and identity theft, and cyberattacks. Fraud detection using AI is more accurate and quicker at the same time as reducing false positives.

8. Cognitive Decision Making

AI that simulates human reasoning, perception and problem-solving capabilities with the aim to solve complex decision problems in uncertain or dynamic environments are described as cognitive decision making. They are the systems that marry up data processing with the bits of cognitive science that can help to reason about context, reason about possibilities, and adapt to what they learn. They are especially valuable in autonomous technologies like self-driving vehicles, robotic process automation, and AI-based legal or medical advisors. By replicating human-like thinking, AI enhances decision-making in scenarios that require judgment, experience, and intuition.

Predictive analytics, NLP and the decision support systems that employ AI in processing of data and similar human reasoning allows for organizations to make smarter, faster, more personalized decisions. Using these tools boosts efficiency, eliminates risk, and creates industry growth in finance, healthcare, logistics and others.

2.3 AI in Supply Chain Optimization and Inventory Management

Supply chain management and inventory optimization is becoming increasingly efficient, less expensive, and more effective with the help of the artificial Intelligence (AI). However, traditional supply chain experiences problems like demand fluctuation, inventory shortage and logistical operation inefficiency. AI based ERP systems resolves the problems related to these issues by virtue of machine learning, predictive analytics, and real time data processing in terms of procurement, inventory levels and distribution strategies. Demand forecasting through AI, driven by historical sales data and market conditions is logically how to predict demand more accurately avoiding stock out as well as stock out. Also, AI driven automation reduces the workload by automating logistic management by ensuring storage, keeping track of the shipments in real time and makes order fulfillment precise. With AI, route planning and delivery scheduling in logistics is more enhanced and is more likely to cut transportation costs and extend service reliability. With the increasing need of supply chain and inventory management, an AI integration can ensure an increase in agility, and therefore also an increase in the minimization of waste as well as an increase in customer satisfaction in the market (Koushik, 2024).

AI in supply chain

The planning, production, management, and optimization of supply chain operations can all be completely transformed by the application of artificial intelligence (AI). AI can enhance supply chain decision-making and operational efficiency by analyzing enormous volumes of data, forecasting trends, and carrying out intricate operations in real time.

This technology has been more well-known recently as new developments like chatbots and generative AI have taken hold and demonstrated the systems' potential benefits for supply chain management. In the meanwhile, the COVID-19 pandemic demonstrated the vulnerability of the global supply chain and the need for improved management systems.

Machine learning (ML), a subset of artificial intelligence (AI), is the process by which a system learns from data sets rather than being preprogrammed with preprogrammed instructions. ML is capable of far more than traditional software. It can estimate consumer demand, identify trends, predict the market, decipher written and spoken language, and evaluate a wide range of variables that might improve the efficiency of a supply chain. There are now more usage cases than ever before.

Adopting AI is vital, but before implementing a new system in a supply chain, it's crucial to comprehend all of its advantages and potential drawbacks. In addition to realizing that such a large-scale optimisation may require time and resources, manufacturers and logistics companies should take the required actions to get their supply chains ready for AI systems.

AI in supply chain work

- AI-powered supply chain technologies are assisting businesses with route optimization, workflow optimization, procurement enhancement, shortfall reduction, and end-to-end task automation.
- Especially for manufacturers who frequently depend on their partners to distribute their goods in a timely and organized manner, a supply chain can become complex. Unlike a standard non-AI system, artificial intelligence (AI) can identify patterns and linkages that help keep all the components of a supply chain in balance. From the warehouse to the cargo freighters to the distribution centres, these patterns can aid in the optimization of logistics networks.
- In order to prevent needless interruptions, modern supply chains must be closely monitored due to their size. Demand planning and the ability to forecast production and warehouse capacity based on customer demand are two examples of how AI systems can help in forecasting. Some are employing artificial intelligence (AI) to extract insights from a larger dataset gathered from Internet of Things (IoT) devices placed across the supply chain.
- Supply chain operations can also employ AI to monitor market trends and inventory levels. AI can improve supply chain visibility, automate physical products documentation, and intelligently insert data whenever objects change hands in inventory management.

Benefits of AI in supply chains

Artificial Intelligence (AI) has made it possible for businesses to increase the rate of operations with data and improve the decision-making process. It reads and processes mega data sets in real time for the purpose of discovery of patterns, trends and correlations which are essential for strategic decisions. Secondly, it reduces drastically the reliance on intuition and manual judgment, thereby making decision making more objective and reliable. This is further enhanced by AI combined with predictive analytics using which market trends, customer needs, and financial risks of their business are forecasted using both historical and real time data. This makes businesses take a proactive approach against challenges and capitalize on opportunities early, thus providing a strong competitive advantage.

Apart from the strategic decisions, AI can help in making the operations more efficient by smart allocation of resources and management of supply chain. It provides for precise budgeting, workforce planning with optimized budget, and the promotion of maximum return on investment by reducing waste. AI supported ERP enables cash flow analysis and prediction, the basis of budget recommendations and revenue growth forecast. When it comes to supply chain operations, AI anticipates demand fluctuations, manages adequate inventory levels and automates procurement using real data on supply performance and market trends. Furthermore, AI-enabled CRM tools enable the organization to track the customers' behavior and feedback so that the services can be delivered to them according to their preferences, optimize marketing strategies, and increase customer loyalty.

Also, AI is essential to the process of routine decision making and organizational support function. It looks for such fraud and risks through transaction patterns to limit financial losses and adhere to regulations and internal policies. AI helps in talent acquisition, evaluates employee's performance and predicts turnover, as well as provides data backed retention strategies in human resources. AI driven real time dashboards give a continuous KPI's insights to leaders on what will happen next and when, so that they are able to make right and timely decisions. AI helps human talent put their attention to high-value, strategic initiatives by automating repetitive work like pricing, restocking, and workflow approvals so that people can become more productive and innovative.

AI in Inventory Management

Inventory management is becoming more and more of an artificial intelligence (AI). However, traditional inventory management based on manual tracking, and rule-based systems are known to overstock, stock out, as well as inefficient operations. AI ERP systems are fed by machine learning, predictive analysis, and real time scans of data to optimize stock levels up, to decrease holding costs and improve order fulfilment. The demand of product forecasting, driven by AI, analyses the historical sales data, market trend, and the external factors which include seasonal and economic conditions and tries to predict the inventory requirement accurately. Also, AI does stock replenishment by issuing restocking orders when inventory reaches predefined thresholds, thereby maintaining the right amount of stock with minimum storage costs. The AI powered inventory systems also have a capability to communicate with IoT sensors and RFID technology, thus tracking the stock movement in real time, reducing losses due to theft, spoilage or mismanagement. Moving inventory management onto AI will boost the efficiency of supply chain, cut waste, and work to make customers pleased via swifter and better order handling.

Inventory management is a key area where AI can bring a lot of benefits.

1. **Demand Forecasting Accuracy:** AI using the **market trends, historical sales data, and outside variables** (seasonality, economy shifts) to predict next inventory requirement which will reduce the overstock and stockouts.
2. **Automated Stock Replenishment:** By using AI driven ERP systems, automatic restocking happens when the inventory comes down to preset thresholds thus ensuring optimal stock levels without intervention from humans.

3. **IoT sensors and RFID technology:** AI can apply to real-time inventory tracking, detecting shortage and minimizing stock losses due to theft, spoilage, and wrong management. AI aids businesses to decrease their storage cost by keeping the inventory at the optimum level, preventing the accumulation of excess stock, and minimizing wastage.
4. **AI Systems bolsters Supply Chain Efficiency:** AI powered systems forecast on the procurement, storage and distribution provides anticipation of delays, detection of bottlenecks as well as recommend secondary or alternative suppliers / routes.
5. **Dynamic pricing and stock optimization:** AI keeps the pricing in line with fluctuations of demand and the market conditions to ensure better profitability and inventory turnover.
6. **Warehousing Predictive Maintenance:** AI Supervises warehouse conditions (temperature, humidity, performance of equipment) to improve capability to prevent damage of the inventory storage.
7. **Real Time Audits:** Automated Inventory Audits are real time which means they reconcile stock records with actual inventory and reduces discrepancies, fraud.
8. **Real time insights and recommendations:** By leveraging AI, businesses having real insights and recommendations which boost the decision making, and it could help them stay ahead of the supply chain disruptions.

Supply chain management is being redefined by AI through its providing of very accurate demand forecasting, inventory optimization, scraping logistics automation, furthermore reducing and eliminating every possible operational inefficiency through data analytics and machine learning. Through this integration, agility, waste reduction, cost reduction, and better customer satisfaction occur throughout the supply chain.

2.3.1 AI-driven demand forecasting and procurement optimization

Demand forecasting and procurement optimization is now being revolutionized by Artificial Intelligence (AI) and is becoming possible to with high accuracy in prediction and through data-driven purchasing decisions. The traditional forecasting methods are based on historical data, manual analysis and these may result in inefficiencies, over stocking or stock shortage. On the other hand, AI based systems use machine learning algorithms, real time market information, sales patterns, customer behavior, seasonal trends and all external factors like economic changes or weather changes. This enables businesses to see the demand fluctuations ahead of time and adjust their inventory level based on that. Automation of supplier selection, negotiations of better terms and cost reduction opportunities are all part of AI's procurement optimization. Combining AI powered demand forecasting and procurement optimization into the ERP system helps in lowering the procurement costs, reducing wastage, improving supply chain efficiency and making sure that the right products are available at the right time to ensure customers delight or profit respectively (Mandl, 2023).

AI-enabled demand forecasting

Traditionally, in order to forecast future inventory needs, "demand driven" approaches have relied on order data. Nevertheless, they can fail in the current volatile market. The challenges? The problems are overstocking when demand falls back down and understocking if demand suddenly jumps. These can put strain on the resources and be a financial burden for businesses.

Enter AI-enabled demand forecasting, a game-changing approach that combines technology with strategy to improve logistical operations. But what is AI in demand forecasting, and how does it transform businesses.

Artificial intelligence (AI) is used in demand forecasting to improve and streamline the forecasting procedure. It takes into account social media activity, online reviews, current market trends, and other information in addition to historical sales data. an integrated strategy that, in contrast to traditional approaches, not only gathers data but also analyses it, finds patterns in it, anticipates changes, and provides predictive insights.

Furthermore, the numbers undoubtedly speak for themselves. AI-enhanced forecasting lowers supply chain network mistakes by 30 to 50%, claims McKinsey Digital. Up to 65% fewer sales are lost as a result of out-of-stock scenarios as a result of the improved accuracy. It is anticipated that supply chain administration costs will drop by 25–40%, while transportation and warehousing costs will drop by 5–10%. With an estimated $1.2 trillion to $2 trillion added to industrial and supply chain planning, the wider economic impact is significant.

These advantages are increased by incorporating machine learning into supply chain management in addition to AI. The real-time insights machine learning offers surpass conventional models that only use historical data, such as exponential smoothing and ARIMA (Auto Regressive Integrated Moving Average). With every new data set, these algorithms adjust as they actively learn from the data. They produce more precise and nuanced projections by taking into account variables like demography and social media trends.

The ability of machine learning to address the difficulties presented by new items is a noteworthy benefit in demand forecasting. The absence of previous data causes traditional models to falter in this situation. Machine learning gets around this, though, by identifying comparable items from the past, examining their lifecycles, and utilising that data to forecast the future of the new product.

Demand forecasting with AI is a philosophical shift rather than merely a technical advancement. Businesses can address the complexity of the modern supply chain by fusing strategic foresight with AI and machine learning. This is the gold standard for inventory management and more of the future.

AI for demand forecasting work

By incorporating AI to this demand forecasting, we are transforming the traditional approaches by utilizing advanced Large Language Models (LLMs) and linking them to an organization's immense datasets. This advanced approach allows users to analyze and delve deeper into data to come up with a more accurate forecast and make the most out of this data to for example stock more optimally, optimize their supply chain or get them better in sync with the market trends.

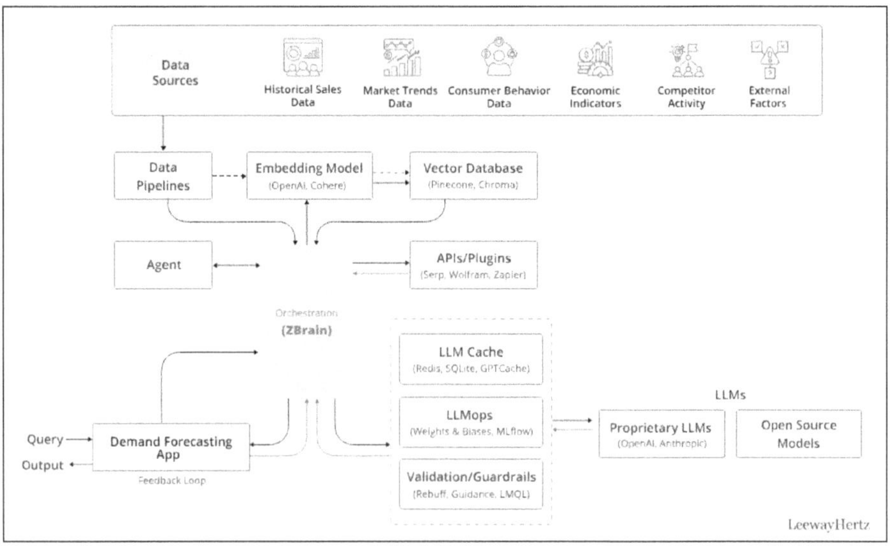

Figure 2.4 *AI for demand forecasting work.*

Source: (Takyar, 2024)

This architecture leverages various components to streamline the AI-driven demand forecasting process. Here's a step-by-step breakdown of how it works:

AI in Procurement

Despite being one of the most popular topics in technology, artificial intelligence is still a relatively new area of opportunity for procurement.

AI is frequently touted as a revolutionary new way to solve issues. All too frequently, discussions focus on hypothetical futures rather than current business realities. Which AI-related procurement opportunities are actually present? How AI can assist in resolving some of the most significant procurement-related issues.

It can sort through the AI jargon and hype with the aid of this guide. Discover the true commercial potential in AI right now. Determine the effects of AI and machine learning on strategic sourcing and procurement. For procurement professionals who are unfamiliar with AI, we provide a wealth of definitions and examples.

Automation in Procurement

Automating your procurement procedures will increase productivity and cut down on cycle time. Employees are released from labor-intensive, laborious, and repetitive duties via automation, which also expedites the process overall.

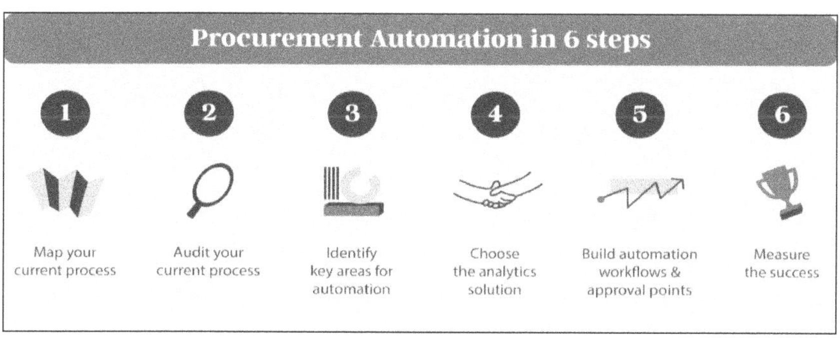

Figure 2.5 *Steps of Automation in Procurement.*

Source: (SIEVO, 2024)

1. Map Your Current Procurement Process

First of all, write down your existing procurement workflow from beginning to end. All this includes from finding needs and sourcing suppliers to purchase order creation, approval and money. Mapping the process allows one to visualize how procurement tasks flow through the organization and how different roles, responsibilities and interdependencies are. It provides a comprehensive understanding into what constitutes inefficiencies and what opportunities for improvement there are.

2. Audit the Current Procurement Process

After mapping the process, audit it in detail in terms of the efficiency, compliance, and its effectiveness. It refers to the analysis of the procurement cycle times, error rates, costs, vendor performance, and the adherence to the policy. The audit enables the identification of gaps, redundancies and bottlenecks that impede productivity or result to overspending. Procurement staff and stakeholders are gathered for input into a well-rounded evaluation and for the purpose of meaningful automation.

3. Identify Key Areas for Automation (Labor-Intensive, Repetitive Bottlenecks)

After audits are completed, point out the most tedious and time-consuming activities in the procurement process. Examples of these could be manual data entry, invoice matching, purchase order approval, vendor onboarding. The most useful form of automation occurs when the tasks they automate are rule based and prone to human error. It helps identifying where to apply automation first to get the most impact on speed, accuracy and resource utilization.

4. Choose the Analytics Solution That Supports Automation

Choose a data analytics platform or procurement software that will accompany your automation. The solution should have the desired feature such as real time dashboards, predictive analytics, AI powered decision support and the integration with existing enterprise systems. It should help users to track the key metrics, identify trends and help them with smart automation workflow. The right chosen platform not only enables automation, but also helps with a better strategic procurement decision making.

5. Build Automation Workflows and Approval Points

Design and build digital workflows that are analogous and superior to the existing procurement processes. It encompasses all this; including automating the requisition approvals, invoice processing and even order confirmations. Decide in advance approve points and problem escape routes so that control and responsibility can be maintained. Automating processes should not be a substitute for the human oversight, but rather automated to reduce the time as well as allow the required checks and balances to be in place.

6. Measure Automation Success and Improve

So, follow up of implementation of automation with the performance of implementation by continuously measuring using some key procurement metrics, like cycle time reduction, cost savings, error rates and supplier responsiveness. Get users and stakeholders' inputs to know areas that have not yet received attention. With these insights, you should use these to fine tune workflows, try to close any gaps, and scale automation to other procurement functions. The automation strategy evolves with the business needs; and this is ensured by continuous improvement.

2.3.2 AI-enhanced warehouse automation and logistics

The artificial intelligence enabled warehouse automation and logistics are a significant turnaround with respect supply chain management, utilizing artificial intelligence to organize processes, lower the cost and never neglect to boost proficiency. Consequently, the use or augmentation of AI driven technologies are replacing or supplementing traditional warehousing processes, which typically rely on manual labor and legacy systems. These advancements make it possible to make real time decisions, do predictive analysis as well as optimize warehouse workflow such as inventory management, picking, packing and distribution. This means that businesses can do more effectively with speed and accuracy to respond to the changing marketplace (Paul Daugherty, 2023).

Figure 2.6 *AI Warehouse Automation: The Future of Logistics Efficiency.*

Source: (INTERROLL, 2024)

Among the most important areas where AI can be used in warehouse automation is intelligent inventory management. By giving AI algorithms examples of what demand patterns look like for those categories of items, it tells it to predict future demand patterns for the items and track stock levels in real time, and decide for itself whether to reorder the items by reordering the items as soon as a threshold is met to reduce the risk of stockouts or overstocking. Stock identification and tracking in the machine vision systems and smart sensors brings more accuracy and leads to less errors in the storage and retrieval of them. AI also assists in warehouse layout optimization based on a usage data and fast-moving goods positioning strategically in order to meet the usage data and speed for access and better utilization of the space.

AI has become important in Logistics and we can say that it's beneficial in route optimization, fleet management and delivering schedule etc. Traffic conditions, weather patterns, combined with delivery constraints are considered by machine learning models then recommending the most efficient route such that fuel usage is minimized and delivery are made on time. Proactive maintenance of the transport assets is also possible using predictive analytics, which reduce downtime and are cost effective. In addition, chatbots and virtual assistants based on AI can help in increasing communication and coordination among logistics teams, so that transparency and customer service are increased.

Also, the integration of AI into warehouse and logistics operation makes collaboration smooth within the supply chain. Airship's ability to connect AI systems with both ERP and TMS allows companies the ability to gain end to end visibility and control. By this holistic approach the operational performance is improved and the strategic text is supported by data driven insights. Together with constantly developing technologies of AI, its role in warehouse automation and logistics will only grow, causing future innovations, sustainability, and as a source of competitive advantage in modern supply chains.

Logistics today is a matter of necessity, not science fiction: automation. AI is one of the transformative technologies that are reshaping the industry in the game changing way. By incorporating AI in warehouse automation, businesses enable unprecedented efficiency, sustainability and scalability of the material handling core and positioning themselves for a more efficient and future-proofed one.

Emerging Trends in AI Warehouse Automation:

This aligns with the broader trends affecting logistics that give rise to warehouse automation:

1. **Sustainability:** Energy-efficient solutions continue to be highly sought-after by both consumers and businesses. AI applications enhance technologies like zero pressure accumulation (ZPA) to maximize energy use and significantly reduce waste. This not only lowers operating expenses but also helps businesses meet their extremely strict sustainability targets.
2. **Plug-and-Play Systems:** The ease of use of AI greatly facilitates the adoption of modular and scalable solutions. The control units, for instance, are preprogrammed, meeting the need to incorporate new parts into the current system in a seamless manner. It facilitates installation time savings and encourages system adaptability to changing operational requirements.
3. **Flexible AGV Integration:** We are improving our AGVs and AMRs by automating them to bridge gaps in short-term or intricate workflows. In order to maximise value in dynamic situations, these devices use artificial intelligence (AI) to ensure that they can cooperate as a component of larger automated systems.

Benefits of AI Warehouse Automation

Warehouse operation is being transformed by AI, and AI has never been more advantageous in moving warehouse operations that way. The most immediate benefits include efficiency gains as AI powered systems work continuously with no limitations of human labor and process and decide faster and accurately. Another must have is the improved safety by eliminating the human intervention in possible unsafe environments, in high-risk warehouses being able to reduce considerably the rate of workplace accidents. Furthermore,

AI contributes to future proofing your business as it avails you to change with the market dynamics, scale its operation efficiently, and seamlessly integrate the new technologies into the system without causing too much disturbance. Finally, AI can help companies lead in terms of sustainability as it enables companies to optimize energy and resource usage, minimize waste, also helping companies meet consumers' and base lines for greener operations, which will be in line with regulatory agendas.

2.4 Challenges and Best Practices for AI-ERP Integration

Systems for enterprise resource planning (ERP) and artificial intelligence (AI) combine to change the fate of the enterprise digital management. For a long time now ERP platforms have been the digital backbone of organizations that served as focal point for organising major functions such as finance, supply chain, human resources and customer relationship management. However, AI is developing as a game changing new force, which can analyze mass of data, detect patterns, automatize tasks, and give predictive insight to make smarter life better and faster (Choudhuri, 2024).

Intelligent AI integration not only improves and changes the workflow of ERP systems but also brings them up from a mere repository of static data to become the dynamic and intelligent engine that drives real-time business optimization. The most important contribution of AI is increasing the accuracy of forecasting to a point where it will bring substantial gains in forecasting accuracy, operational efficiency, customer personalization, and strategic agility. With AI taking over routine back-office tasks as well as helping in predictive analytics for demand planning, organizations will be more proactive, adaptive and competitive as the world becomes more data-driven.

But despite this, realizing the maximum potential of AI ERP integration is not an easy task. Typically, the way towards intelligent

ERP transformation involves a mixture of challenges, including data fragmentation, outdated infrastructure, organizational resistance, and fears of data privacy and compliance. If these obstacles are not properly addressed, adoption can be restricted, ROI can be poor and maybe even a COC (AI initiative) can be derailed.

That is why, before integrating your business with these firms, you must know the common pitfalls and proven strategies that lead to success. In this section, the challenges faced while integrating AI and ERP are discussed, and best practices that the organizations can follow to overcome these challenges are outlined. An integration strategy mindset with which companies tackle the integration of ERP ecosystems will enable them to build resilient, intelligent ecosystems that support continuous innovation and long-term growth. ERP systems integration with AI is a very complex challenge that requires much thought and execution, but at the same time has many opportunities for integration.

Modern business has great transformative opportunities that can be built in Artificial Intelligence using Enterprise Resource Planning (ERP) systems. AI can take ERP platform to enterprise ecosystem by elevating them to intelligent predictive analytics, intelligent automation, real time decision making, resource optimization. Said road is not smooth however, with technical, organizational and strategic challenges. These barriers have to be anticipated by businesses who must adopt best practices to use AI's full potential within their ERP environments. It is the most prominent challenges, and this section discusses about the most comprehensive strategy to overcome them.

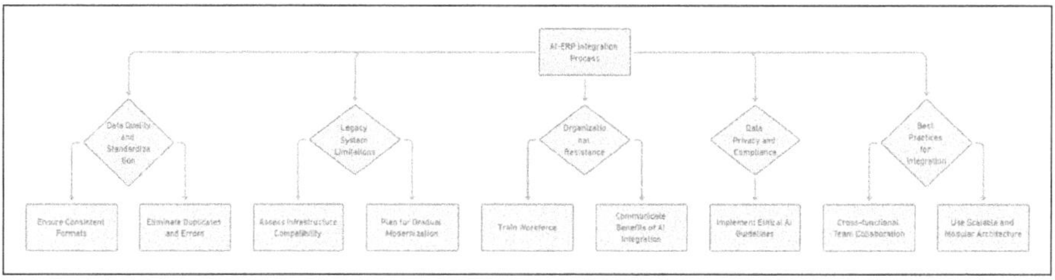

Figure 2.7 *AI ERP integration process.*

Source: (Self Generated)

1. Data Quality and Standardization

Data quality is one of the greatest challenges for the integration of AI with ERP systems. Clean data will bring them on to their knees. Unfortunately, there are many organizations today that are struggling with data silos, different formats of data, outdated records, and copying entries in their ERP server. Poor data quality heavily undermines the performance of AI models and is bad news for getting accurate insights and unfortunate predictions.

To avoid this, organizations may need to spend some time in data standardization and cleansing before including AI components. Such as owning the data with defined roles and then audited on a regular basis with the use of data quality tools helps create a basis for the solid foundation. Besides, it is necessary to unify the data across different departments to break the silos so that the AI algorithms will be given a complete picture of the enterprise for their more accurate analysis.

2. Legacy System Limitations and Infrastructure Compatibility

Most organizations continue to use legacy ERP systems built a decade or two before the days of mainstream AI came to be. And then, in these systems, there is not the technical flexibility, computational capacity, or modern architecture that it takes to support AI. Such interrace obstructions between 'old' infrastructure and new generation of

AI solutions may fail to implement or to reap maximum benefits of automation and intelligence due to incompatibility.

Organizations must prepare themselves to get through this hurdle by looking to phased modernization strategies. Cloud based ERP platforms are more flexible and can be more easily integrated with AI modules through APIs and plug in. If not, middleware solutions can bridge the gap between new and old technology to let AI have safe contact with legacy systems. Practical pathway for many enterprises is a hybrid infrastructure to gradually replace obsolete elements retaining the most important operations.

3. Organizational Resistance and Workforce Readiness

Transforming from a technological shift to an organizational and cultural shift, in a way the integration of AI to ERP is not at all. Fear of automation, the threat of job (or enjoyment of specific tasks) from new AI processes, or the ignorance of the types of processes may deter the most loyal employees from embracing new types of processes. The resistance can slow down adoption, lower productivity during the transition, and in general contribute to the failure of the integration.

The solution to these calls for a change management strategy that is proactive. In order to foster an environment to learn and innovate, leadership is required. The first of which is to invest in continuous training and development programs to increase AI literacy. The sooner AI is introduced into the process, the sooner employees can be engaged and understand how AI will enhance their roles, rather than replace them. By accelerating the changes, internal momentum and smoother transition will be helped by cross-functional collaboration and change champions of the departments involved.

4. Data Privacy, Ethics, and Regulatory Compliance

Security issues, safety, and compliance of privacy, where sensitive financial and operational and employee data because of AI systems that are processed in ERPs systems is critical. The more automation in scale and intelligence, the higher is the risk of a data breach or misuse of information. Additionally, all the integration efforts in industries have to observe strict regulations such as GDPR, HIPAA among others which include regional laws in particular.

That being said, businesses have to have an all-encompassing security protocol in place which includes encryption, access controls, data anonymization, and secure API gateways. To ensure legal, ethical and compliance frameworks are aligned in the blend from the start. The design and deployment phases of AI-ERP systems could include involvement of cybersecurity experts, compliance officer and, legal advisor to mitigate risks and ensure accountability. Transparent data practice and regular audit helps increase the level of trust and secures enterprise data.

5. Best Practices for Seamless AI-ERP Integration

In order to overcome these challenges and successfully integrate AI into ERP systems, organizations have to put best practices in place that are based on business goals, technology alignment, operational agility.

But starting on a project with small scale pilot project that focus on high impact, low risk process like automation of invoice processing, demand forecasting and inventory optimization is the recommended approach. In fact, these use cases give rise to quick wins, to validate the AI models and build stakeholder confidence. If proven, the integration can be taken across larger ERP modules like supply chain, finance, human resources, and customer relationship management.

The right AI tools need to be selected. Therefore, businesses should consider solutions that are in line with their existing ERP vendor ecosystem or easily support seamless integration via flexible APIs and micro services. Customization for deployment is accelerated using platforms with open AI models or connectors that are fully built.

In decision critical areas, the AI loop should always include human oversight. Data analysis and performing repetitive tasks is something that can be handled by AI, but only human judgment can interpret outputs, manage exceptions and make ethical decisions.

Last but not least, a subsequent monitoring of the post integration is necessary for a sustained success. Metrics for AI performance should be set up by organizations such as accuracy rates, time savings, cost savings, etc., regarding user adoption levels. The AI ERP system receives continuous feedback loops, iterative model improvements and adaptive learning so that it evolves with the business continuously and stays relevant when the environment is dynamic.

2.4.1 Overcoming implementation barriers in AI-ERP

Artificial Intelligence (AI) has been fast changing the operational landscape of modern enterprises by being integrated into Enterprise Resource Planning (ERP) systems. Traditional ERP is only for basic data processing and decision support, AI stands ERP in the sense of advanced analytics, intelligent automation and real-time decision-making capabilities. It is very complicated to apply AI in ERP frameworks, however. When an organization has to integrate, the process is not a simple one, especially if the challenges are numerous and retard or slow down the process. There are technical and organizational barriers that may not be interoperable systems, or may resist change from current systems (Moallem, 2021).

The main hurdle in AI ERP implementation is the data related piece. The best fuel for such AI systems is large amount of high quality high structured data, but most organizations have fragmented or siloed data across departments. Then data consistency, old databases and lack of a centralized data governance policy can hinder the training and effectiveness of AI models. Moreover, to properly use AI tools in existing ERP workflows without adversely affecting the ongoing activities, the contextualization, planning, and technical expertise required amount to a lot, and most organizations may not be able to provide it internally.

Another major barrier is the human and organizational aspect. That's because employees are afraid AI will replace their role and so they resist and adopt at low rates. Secondly, it does not have internal AI knowledge and sufficient personnel to make and deploy AI systems in ERP platforms. Important things to do to have a culture of acceptance, change management, training and communication about the role of AI as an enabler, not a replacer. Secondly, they should work towards aligning AI with business goals, and in facilitating a cross functional collaboration to have a single business.

However, these barriers must be overcome and this is best achieved through a phased and strategic implementation approach by organizations. Additionally, it includes infrastructure upgrades funding, setting up robust data governance practices, and working with the AI ERP solution vendors that offer flexible and customizable tools. All equally important, besides upskilling, hiring, and creating an innovation culture. By having a roadmap of how your enterprise will achieve value first and your journey thereafter, you can stakeholder buy in and avoid focusing on quick wins for quick wins' sake, and enter the AI ERP complexity space successfully, and ultimately leverage it for your enterprise.

2.4.2 Optimizing AI for enterprise-wide digital transformation

Digital Transformation of an enterprise with AI is about bringing artificial intelligence capabilities in sync to the organization's strategic objective, innovation and operational efficiency. It means that AI is not built into one place, but rather a complete lifecycle approach to the business across supply chain to production to marketing and to customer service. Such transformation's success hinges on the ability to integrate AI with existing business enterprise resource planning (ERP) systems and continue to flow data and decision-making insights. AI can not only automate routine tasks, but also do predictive analytics that will help the enterprises to predict changes in the market thereby manage risks proactively and respond to dynamic business needs with agility (Shemshaki, 2024).

Unlike traditional automation, AI brings adaptability through learning, contextual awareness, the ability to learn to do the job in future, and leverage of historical data. Artificial narrow intelligence (ANI) is widely used in the manufacturing and engineering settings to solve a specific task, for example, computer aided design (CAD) optimization, predictive maintenance. AI algorithms are different from rigid rule-based systems in which data can be processed in real time and how changes are unexpected can adapt to changes and also the self-correction also happens over time. One example of this is that generative design with AI in CAD applications allows engineers to input parameters and get many optimized design alternatives in significantly shortened development cycles and boosted innovation.

With concentrations of digital transformation raw data, the key is to turn raw data into actionable intelligence using AI. For instance, computer vision AI is used in augmented reality (AR) applications for example platform such as Vuforia, that recognizes physical components, to improve the user experience and accuracy of the app. On the other hand, taking the example of Thing Worx, predictive analytics are used by systems to check the performance of the asset and forecast maintenance requirement for reduced downtime and optimal allocation of resources. These examples show how AI can become the 'brain' for such a set of tools that together enable smarter, faster and more informed decision making across the enterprise.

But for enterprises that choose to back away from transforming into a digital enterprise with the inclusion of AI, they run the risk of fast dying in a growing, competitive, ever accelerating market. First of all, AI not only enhances efficiency but also enables organizations to get long term scalability, sustainability, and customer centric innovation. To enable this process, it is important to cultivate an AI-literate workforce, scale infrastructure, implement a culture of experimentation and constant improvement. If the predominant role of AI is embedded at the very outset of digital transformation, new value streams can be created which will gain enterprises a mature positioning in the digital economy.

Multiple Choice Questions (MCQs)

1. What is the primary goal of integrating AI into ERP systems?

A. Replacing all human roles with robots

B. Enhancing business operations through automation and data-driven insights

C. Increasing product prices

D. Creating new ERP software from scratch

2. Which of the following best defines Intelligent Process Automation (IPA) in the context of ERP?

A. Manual entry of business data

B. Basic task scheduling features

C. The combination of AI and automation to streamline business processes

D. Cloud storage of ERP data

3. How does AI-powered decision support benefit business operations?

A. By eliminating the need for all human decisions

B. By providing real-time insights and recommendations based on data patterns

C. By automating payroll systems only

D. By slowing down decisions to ensure accuracy

4. What is a major benefit of AI in supply chain optimization?

A. Eliminating the need for supply chain managers

B. Reducing demand for goods

C. Improving demand forecasting and procurement accuracy

D. Replacing transportation systems

5. Which of the following uses AI to improve warehouse operations and logistics?

A. Manual inventory tracking

B. AI-enhanced warehouse automation

C. Paper-based supply chain management

D. GPS-based employee monitoring

6. What is a common challenge faced during AI-ERP integration?

A. Excessive availability of AI expertise

B. Difficulty in collecting business data

C. High data quality and seamless integration with legacy systems

D. Lack of software licensing requirements

7. Which of the following is a best practice for optimizing AI in enterprise-wide digital transformation?

A. Avoiding employee training

B. Isolating AI systems from ERP

C. Promoting AI literacy and aligning AI initiatives with business goals

D. Using AI only in R&D departments

8. AI-powered ERP systems can help reduce operational costs by:

A. Adding more manual checks

B. Increasing redundancy

C. Automating repetitive tasks and optimizing resources

D. Requiring more human approvals

9. What makes AI-enhanced ERP systems better suited for dynamic market conditions?

A. Their rigid rule-based processing

B. Their reliance on manual data entry

C. Their ability to learn, adapt, and make predictive decisions

D. Their lack of data dependency

10. Which AI capability supports ERP systems in generating optimal product or service recommendations?

A. Image recognition

B. Machine learning and predictive analytics

C. File compression

D. Virtual reality

Answers

1	2	3	4	5	6	7	8	9	10
B	C	B	C	B	C	C	C	C	B

Chapter 03

AI in Risk Management and Compliance

3.1 Introduction to AI in Risk Management

Risk management is an important process in most business sectors like the finance sector, healthcare sector, cybersecurity sector, supply chain management and many more. This includes the identification, evaluation, and reduction of potential threats that could disrupt the operation of business, the stability of financial affairs or the full abidance to the regulations. Historical-Based approaches, rule-based approaches, and the expert judgements are all heavily used in traditional risk management methods. While modern risks are becoming more complex (cyber threats, economic volatility; operational uncertainties), Artificial Intelligence (AI) has proved to be a very powerful tool to strengthen the risk management strategies (Data et al., 2016).

Machine learning, predictive analytics, and natural language processing (NLP) are used to implement AI in risk management system by analyzing huge volume, structured and unstructured data. Rather, these models are AI driven which means, that it helps you to detect patterns, identify anomalies and reveal potential risks within the real time. However, unlike in the convention's manner, AI can process it all from multiple data sources, like financial reports, social media trends, news feeds, and regulatory updates to solve the risk prediction before they become a reality. This proactive approach plays the role in reducing losses in the event of an incident, in making better decisions, and taking care of regulatory compliance.

Fraud detection represented one of the key apps in which AI is being used for risk management, alongside cybersecurity. Continual buying pattern analysis and other activity such as transactions, network activity and user behavior can be continuously monitored by the AI powered systems so as to detect suspicious patterns and thwart fraudulent activities real-time. For example, in financial services, AI models evaluate credit risk based on customer's behavior, transaction history, as well as economic indicators to improve lending decisions (such as whether or not to lend money to a customer) for banks. In supply chain risk management, similarly, with the help of AI, it predicts a disruption based on geopolitical risks, weather patterns and supplier reliability.

This AI-driven risk management has also some problems. If data privacy, model interpretability, or bias in the AI algorithms is a problem, then they have to be worked upon prior to implementation. The question is whether organizations can keep things transparent and accountable without jazzing into that excess of automation.

Why do Companies Need AI in Risk Management?

Nowadays, companies face many risks which is why having AI is an essential tool to identify, assess, and mitigate potential threats with speed and accuracy.

1. **Security Risk:** Security risks arise when AI systems have vulnerabilities that attackers could exploit or users might unknowingly introduce. It leads to unintended outputs. These risks can involve attacks targeting the model itself, the data it processes, or the underlying software. Whether proprietary, commercial, or open-source, AI and ML in data integration models face threats like supply chain risks, data poisoning, prompt injection, personal information leaks, and even theft of the model itself.

2. **Ethical Risk:** Ethical risks occur when an AI system's behavior clashes with societal norms, legal requirements, or governance policies. These issues often stem from biases in training data or evolve from new patterns in production data over time. Common examples include producing biased predictions, generating toxic or offensive outputs, or responding in ways that display prejudice or exclusion.
3. **Operational Risk:** Operational risks surface when an AI model's predictions deviate from expected outcomes. These issues might be caused by phenomena such as data drift, hallucinated results, corrupted datasets, unusual inputs, or broken data pipelines. Silent failures pose a unique challenge—they don't cause the model to crash but subtly degrade its performance, making them hard to spot while impacting downstream processes.

How Does AI in Risk Management Help Organizations?

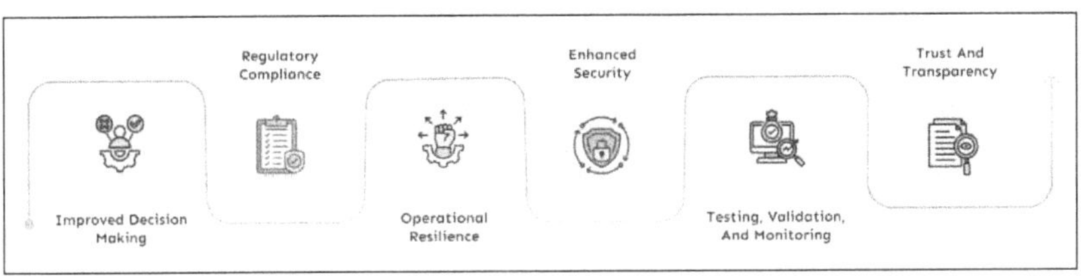

Figure 3.1 *How AI in Risk Management Helps Organizations?*

Source:* *(Lab, 2023)

1. Enhanced Security

AI risk management is great for boosting an organization's security, particularly in the cybersecurity. By regularly assessing risks and conducting audits, organizations can spot potential vulnerabilities throughout the entire AI lifecycle. Once risks are identified, the next step is implementing strategies to address or eliminate them. This could include technical solutions, like strengthening data security or enhancing model robustness, as well as organizational changes, such as creating ethical guidelines and tightening access controls.

Taking a more proactive approach to threat detection and response helps companies tackle risks before they blow up into bigger problems, reducing the chances of data breaches or the impact of cyberattacks.

2. Improved Decision-making

AI risk management also helps organizations make better decisions overall. By blending both qualitative and quantitative data, including statistical analysis and expert opinions, companies get a clearer picture of their potential risks. This well-rounded view allows organizations to prioritize the highest risks and make smarter decisions about deploying AI—striking a balance between innovation and risk control.

3. Regulatory Compliance

With a growing focus on data protection, various global regulations have popped up, such as GDPR, the California Consumer Privacy Act (CCPA), and the EU AI Act. Failing to comply with these laws can result in major fines and legal issues. AI risk management can help companies stay compliant, which is especially important as regulations around AI change just as quickly as the technology itself.

4. Operational Resilience

AI risk management helps organizations keep things running smoothly by quickly addressing potential risks tied to AI systems in real-time. It also promotes long-term sustainability by encouraging clear management practices and methods for using AI responsibly.

5. Increased Trust and Transparency

AI risk management is all about making AI systems more ethical by putting trust and transparency at the forefront. This usually involves bringing together various stakeholders—like executives, developers, data scientists, users, policymakers, and even ethicists. Having such a diverse group ensures AI systems are built and used with responsibility in mind, considering everyone's interests.

6. Ongoing Testing, Validation, and Monitoring

Regular testing and monitoring are crucial to tracking an AI system's performance and catching emerging threats early. This helps organizations stay on top of regulatory requirements and address risks before they turn into bigger problems, ultimately minimizing the impact of potential threats.

An advent of AI into risk management that facilitates a machine learning, predictive analytics and real time data processing-based approach to identifying, evaluating and tackling the risks is a revolution. This improves security, improves decision-making, ensures regulatory compliance, and improves operational resilience so businesses can take pro-active stance on resolving risks for cybersecurity, financial fraud and supply chain disruptions. While the implementation is being considered, there are still data privacy, ethical concern and AI model bias challenges that need to be carefully taken care of to make sure we maintain transparency and accountability. Combining AI driven automation with human oversight allows organizations to find a balanced and responsible approach to risk management that will allow such business to adapt long term and continue to build trust in an ever-changing business environment.

3.2 AI Models for Predictive Risk Assessment

Dependency on predictive analytics has changed the game when it comes to the risk of capital companies, thanks to Artificial Intelligence. However, historical data, the combination of expert judgment and the static rule-based systems used to manage risk were not particularly powerful, particularly against altering threats. Where, however, machine learning, machine learning with deep learning, and the further statistical models to the abovementioned detectors dynamically have vast structured and unstructured data in real time. AI driven models help these models to identify patterns, forecast future risks and make decisions in simpler way. Integrating AI into risk management helps businesses switch from a reactive to a proactive approach to minimize the harm of uncertainties before they arise (Science, 2024).

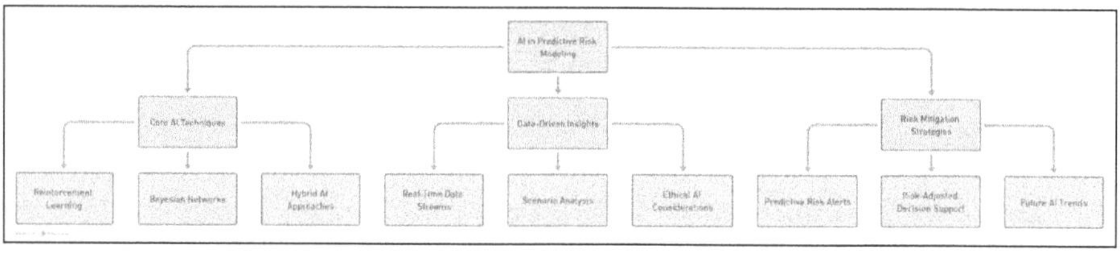

Figure 3.2 *AI Risk Modeling Overview.*

Source: (Self-generated)

1. Core AI Techniques in Predictive Risk Modeling

- **Reinforcement Learning for Adaptive Risk Management**

  Reinforcement learning (RL) is the part of ML that learns through repeated interaction with dynamic environment. Contrary to the fixed rules in conventional methods, RL permits AI systems to change at runtime according to the feedback received from the real world. Such capability is especially useful in situations where the threats are constantly changing. For example, an RL approach will help financial institutions use the changing transaction behavior to dynamically improve their fraud prevention strategies. In the same way, RL powered systems adjust security protocols for fighting off the latest attackers and secure defenses.

- **Bayesian Networks for Uncertainty Quantification**

  Probabilistic models of Bayesian network are used to quantify and manage uncertainties in risk assessment of organizations.

The way these models work is by relying on statistical inference to predict an uncertain situation of risk. For example, in insurance, Bayesian networks try to calculate insurance premiums and probability of claim based on several risk factors (demographic data, medical history, financial status). AI driven Bayesian models can help medical risk assessment with forecasting the likelihood that patients have diseases through genetic and environmental factors, and this can seek early intervention and better patient care. Businesses can integrate Bayesian reasoning to make data driven decisions but in light of uncertainties.

- **Hybrid AI Approaches for Holistic Risk Assessment**

  Risk assessment accuracy can be improved by combining a number of different algorithm techniques that use multiple data sources. In case of ensemble learning, for instance, decision trees, neural networks, and regression models are combined to offer an overall risk assessment. In the financial services industry, this method helps to boost the credit risk scoring based on customer behavior, marketplace trends and economic indicators. Like this, multi-modal AI systems which process text, images and sensor data at the same time are used in security and in predicting disaster risk. Other hybrid approach to AI includes fuzzy reasoning that allows those with imprecise or ambiguous data to be handled. This is particularly useful for areas like operational risk assessment and fraud detection.

2. Data-Driven Insights for Risk Identification

- **Real-time data Streams for Proactive Risk Monitoring**

  Real time AI models consume data from streams of real time data and these streams can be from any source relevant to an organization's risk universe. For instance, the Internet of Things (IoT) accumulates enormous amount of sensor data that AI can analyze to determine a potential industrial machinery failure and minimize time of downtime as well as maintenance costs. AI based sentiment analysis helps the gain insights from social media, news articles and market data and predict stock fluctuation and economic shift in the financial sector. Likewise, AI models are used for prediction of disasters risk through analysis of meteorological data as well as satellite imagery to predict the occurrence of extreme weather events, and counteract the disaster risk, which contribute to their responses to natural disasters.

- **AI-Enabled Scenario Analysis for Strategic Planning**

  Based on AI driven scenario analysis, organizations are able to prepare for the possible risk contingencies. AI is able to simulate various economic downturns and assess the impact financial crises may have on a company's operations and suggest mitigation strategies. In supply chain management, AI too looks at geopolitical risks, supplier dependencies and transportation delays to predict disruptions and suggests dissimilar sourcing strategies. The second important area where AI comes in handy in forecasting policy changes and what change will have on businesses so that they are able to adjust accordingly to these changing legal frameworks.

- **Ethical AI in Risk Assessment: Bias and Fairness Considerations**

  Although AI-driven risk assessment has great benefits, there are also ethical issues such as bias and fairness. Bias in the dataset means bias in risk evaluation because risk evaluation can be used for decisions like credit scoring and insurance underwriting. To be fair with their AI driven risk assessments, organizations have to provide bias detection techniques. In addition, trust and accountability require AI decision making to be transparent as well as explains. Ethical AI frameworks development helps organizations maintaining innovation, fairness and compliance, respectively.

3. AI-Driven Risk Mitigation and Response Strategies

- **Predictive Risk Alert Systems for Crisis Management**

  Predictive alert systems based on AI give organizations a chance to take preemptive action in the event of potential crises. AI in the financial sector makes use of macroeconomic indicators to predict market downfalls and allow for adjusting your portfolio. AI driven threat detection systems are used in cybersecurity wherein AI based systems detect anomalies in network activity and then set up security steps to restrict data breaches. AI is also changing the public health by predicting the spread of diseases and helping governments to allocate resources for epidemic control.

- **Risk-Adjusted Decision Support Systems**

  Risk management is made better by AI powered decision support systems that provide the optimum strategy out of the real time data analysis. In the investment management, AI can assess financial information to recommend portfolio adjustments which aim at balancing risk and return. AI is also used in the construction of corporate risk scoring models that use information on a company's financial status, operational risks, and market position for the benefit of investors and stakeholders in taking appropriate decisions. Furthermore, AI driven operational models of efficiencies favor resource allocation and waste minimized in unfavorable environments among others.

- **The Future of AI in Risk Assessment**

  AI driven risk assessment is about to see a boom in the future. For instance, quantum AI is capable of making a positive impact by speeding up risk simulations for complex variables. Next to develop such self-healing autonomous cybersecurity defenses that self-adapt without human involvement are autonomous AI systems. Furthermore, AI controlled regulatory compliance will allow businesses to adapt to changing legal landscape through a continuous and complete adherence to ideal business practices. The more the AI technology develops, the more it will be used to predict the risk, which will be essential for safeguarding businesses, economies and societies.

3.2.1 AI-driven risk identification and mitigation strategies

Today, it seems that companies are confronted with a sea of risks when they operate in a fast-changing business environment, such as financial unpredictability and cybersecurity breach, business discontinuity and violation to regulatory inconsistency. Current risks are orders of magnitude faster and more complex than what traditional risk management is used to, and these are hindered by using historical data and people's expert opinion. One of the revolutionary solutions in terms of Artificial Intelligence (AI) in which companies can proactively detect the risks, evaluate risks and implement required mitigating measures. Such systems leverage on ML, NLP, predictive analytics and automation, alongside AI to develop and strengthen risk clearances and resilience to threat to the decision making (Karim et al., 2023).

First of all, risk identification is one of the advantages of AI as it can process huge amount of structured as well as unstructured data at the same time. Unlike conventional method AI does not depend on periodic 'reviews' like it does but monitors a number of data sources (financial reports, social media activity, customer feedback, market trends and so on) as well as regulatory updates. With these inputs, AI can find which of them could easily become potential risks already in the early stages: fraud patterns, security vulnerabilities, supply chain interruptions or economic drops. Advanced anomaly detection algorithms allow organizations to detect anomalies and emerging threats ahead of time so as not to expose them and let them grow out of control, and to be proactive instead of reactive about the risk.

Apart from risk identification, AI offers the predictive part and the auto response mechanism. AI powered predictive models with its bases on historical trends and real time data, it can predict the risk probabilities, and this will make organizations to proactively so that risk can be reduced at least. As an example, AI can predict the credit default risk and change the lending policies. AI intrusion Detection System is a cybersecurity tool which is an AI (Artificial Intelligence) tool to detect the suspicious network activity and automatically implement the counter measure to protect the data from being breached. Similarly, in the healthcare AI driven risk assessment models help to figure out the disease outbreaks early on, which then helps authorities to implement containment strategies along with redundancy and allocation of resources.

Mitigation strategies driven by AI are also effective in optimizing resource allocation and risk response strategies in the process of decision-making. Digital twins and scenario analysis models that are created using AI-powered simulations let businesses try different risk scenarios and calibrate various mitigation strategies in advance before deciding to use them in their real world. This predictive capability supports organizations in building additional levels of contingency planning, minimizing the impact on the organizational operation, and increasing the degree of operational resilience. It also reduces human errors and brings down time of responding to risk situations to a speed than what a human can do and implement timely interventions in critical situations by acting on their behalf.

AI driven risk management indeed provides with great benefits as there are no obstacles, but at the same time there are also the challenges regards data privacy concerns and model bias alongside the lack of transparency on the decision taken. To prevent unfair and unethical results, organizations have to put in place the ethical AI frameworks in order to ensure fairness, accountability, and compliance with the regulatory requirements. Businesses can leverage the full potential of AI to identify and mitigate business risk that AI can find by refines the AI models and integrated with human oversight. With the growing AI technology, AI will play a more and more important role in risk management and help the organization to negotiate uncertainties, and create a more secure and robust future.

3.2.2 Machine learning models for operational risk analysis

Operational risk is a big challenge to organizations from all the industries, ranging from system failure to human error, supply chain disruption, regulatory noncompliance, and same the cases of cybersecurity breach. If not managed properly, these risks will have financial losses, reputational damage, legal liabilities and disrupt business operations. Current risk assessment methods rely heavily on past data, human assessment and judgement. These approaches have been effective to some extent but they are not powerful enough to discover emerging risks in real time. Operational risks are extremely dynamic and the industries like finance, healthcare and manufacturing are highly dynamic, and so it requires more sophisticated and proactive risk management solutions (Harapan et al., 2020).

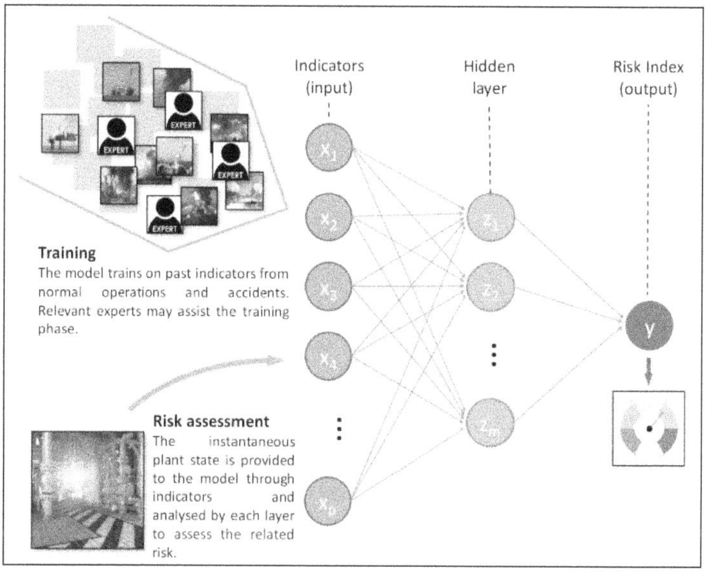

Figure 3.3 *Representation of machine learning approach for risk assessment.*

Source: (Paltrinieria et al., 2019)

Machine learning (ML) has pushed forward the advance of operational risk analysis in terms of making use of data to produce highly automated, highly predictive, and as such, data driven, risk analysis outputs for enabling better risk detection, skill in distinguishing poor versus good outcomes, and risk mitigation strategies. An advantage of ML models is they are able to process massive amounts of accumulated structured and unstructured data, picked up from multiple sources, such as financial reports, operational logs, social media trends, etc. These models include NLP, deep learning approaches, and anomaly detection for finding hidden patterns, uncover vulnerabilities, and predicting impending failures before they worsen into bigger issues. Hybrid systems that use ML power, for example, fraud detection systems ML-powered are made to analyze the transactional data in real time in order to identify suspicious activities, and predictive maintenance models within industrial environment integrate to forecast equipment failures, decreasing downtime as well as maintenance costs.

Despite the immense benefits that ML offers in operational risk analysis, however, a number of problems still exist — for instance, how to ensure data quality, deal with biases in algorithms, and so on — and maintaining model transparency and interpretation. deploying AI driven risk management system also has to take into account of ethical concerns such as data privacy and regulatory compliance. However, automation of human processes without proper governance, lack of control over ML models and little or no human oversight needs to be balanced with accountability. As the AI and ML technologies continue to evolve, impact and the role they are going to play in helping businesses to be more resilient, streamline risk management processes and get ahead of emerging threats in this ever-changing risk landscape is going to be more vital.

Advanced machine learning algorithms, predictive analytics, automation and AI are combined with risk identification and mitigation strategies to detect potential threats in real time and take proactive measures to prevent them from becoming actual threats.

AI-Driven Risk Identification and Mitigation Strategies

AI has sped up the risk finding and risk mitigating process for organizations by allowing them to make proactive decisions quickly, such as when faced with a data deluge. Unlike traditional risk identification approaches that collect historical data and have fixed (historical) models, machine learning (ML), predictive analytics, and anomaly detection (Mathur and Talatar 2011), AI-driven risk identification use prediction to identify threats before they explode. AI models process sophisticated structured and unstructured data from all sorts of sources like financial reports, market trend, social media and cybersecurity log and use such data to identify risk irregular patterns. An approach of trying to be proactive helps organizations take actions in real time in order to avoid losses that may occur in the future. Furthermore, AI enabled risk mitigation strategies enable an automated response through the AI and reduce fraud, cybersecurity threat response and regulatory compliance monitoring. Such AI driven simulations as digital twins allow businesses to play multiple risk scenario simulations, have refined contingency plans and better allocate resources. Nevertheless, AI in risk management can be beneficial in the context of model bias, data privacy concerns and transparency in AI decision making, yet have to be addressed.

Leveraging Machine Learning for Operational Risk Management

Operational risks — such as disruption and cybersecurity threats due to supply chain disruptions, system failures and more — can severely damage the organization's operation and brand reputation. Reflecting on aforementioned risks, machine learning models offer a data driven approach of tackling and detecting these risks by identifying the hidden pattern and predicting future failures. Instead, supervised learning models, such as for example decision tree and neural network, analyze previous occurrences and classify risks as well as suggest preventing actions. An unsupervised learning models such as clustering algorithm and anomaly detection uncover unexpected risks by analyzing out of data. The models in reinforcement learning change with the changing conditions and continually refine risk management strategies. Further, NLP is used in the analysis of regulatory documents, incident reports, and customers feedback to find compliance risk. The integration of machine learning into operational risk analysis allows businesses to improve efficiency, decrease losses and businesses, and be more resilient to uncertainties. Nevertheless, data quality, interpretability and ethical concerns need to be accommodated to realize widespread adoption.

1. AI Models for Predictive Risk Assessment

AI Based predictive risk assessment, which makes organizations forecast probable risks more effectively helping take preemptive action. Traditional risk models use static risk matrices, whereas the predictive analytics of A driven predictive analytics change with changing conditions. Such an algorithm of AI models like regression analysis, deep learning or Bayesian network use huge amount of data and perform prediction of financial, operational and cybersecurity risks. For example, AI is used to determine market trends, credit histories and economic indicators for financial risk management to forecast investment risks. Predictive models in health care study patient data and identify the likelihood of health complications and to make suggestions for preventive care. Supply chain management would not be successful without the use of AI as it can predict delays, supplier failures, and political risks that may disrupt operations. With the help of AI, business owners can foresee risks and absorb disrupts and economic losses. Nevertheless, for AI driven predictive assessments to be reliable, organizations must be sure to maintain the model transparency and address the biases.

2. AI-Powered Decision Support Systems for Risk Management

The problem of risk management is tackled by using a Decision Support Systems (DSS) based on Artificial Intelligence (AI) that provides real time readings and advice based on data driven analysis. These systems, based on research involving, for example, automation involving the machine learning, quantitative big data analytics, and reasoning, help decision makers assess such risk scenarios and select preferred mitigation strategies. With AI, the DSS drives scenarios of risk and tests different responses with the potential outcomes that result from them before the organization takes action. In financial services, Credit risk and fraud likelihood is evaluated with the help of AI enabled DSS to help the banks make informed lending decisions. The cyber security uses AI based threat intelligence platforms to explain the risk levels and the countermeasures to be employed for protecting the sensitive data. Furthermore, AI powered DSS in disaster management assists in better prediction and disposition to the natural disaster. By taking AI into the decision support, business becomes more agile and accurate in risk management and therefore lessens the loss due to uncertainties.

3. Ethical Considerations in AI-Based Risk Management

When AI is deeply embedded in risk management, ethics of fairness, transparency, and accountability become an obligation. The main problem with AI models is their bias: since data used to train the algorithm are skewed, the risk assessment may also be unfair if, for example, credit scoring, hiring process or law enforcement are involved. Another important aspect is ensuring AI explainability so that decision-makers can understand how AI reaches at-risk assessments. Privacy concerns also arise from the fact that much of AI driven risk management relies on hefty amounts of personal and corporate data to be transparent on data security and compliance with regulations like GDPR and CCPA. Thus, to build trust in AI driven risk management, organizations must frame it in ethical AI frameworks, develop robust governance policies and have transparency itself as a major priority. By handling these ethical issues, businesses can use AI for risk mitigation while keeping a good reputation and faith in the public.

3.3 AI-Powered Compliance Monitoring and Regulatory Analysis

Ensuring regulatory compliance, especially regulatory change management, is critical to any business. This entails strict adherence to laws, guidelines, and specifications relevant to business processes (Elliott C. Kulakowski, 2006).

AI significantly eases this burden by automating the monitoring and reporting processes. It can swiftly sift through vast amounts of regulatory content to spot new requirements or updates to manage adherence to regulatory requirements. AI systems can be designed to stay updated with the latest regulatory changes, ensuring businesses remain compliant. Machine learning, a subset of AI, excels in pattern recognition and anomaly detection, which are crucial in compliance.

It can analyze historical data to predict potential compliance risks, enabling proactive risk management. Machine learning algorithms can also learn from new regulatory data, continuously improving their accuracy in identifying compliance issues.

AI plays a crucial role in risk management within compliance. It can predict potential compliance risks and suggest measures to mitigate them. AI systems can monitor real-time transactions, providing alerts on suspicious activities that may indicate non-compliance or fraud. This real-time analysis helps in quick decision-making to prevent compliance breaches.

Benefits of AI Adoption in Compliance

Artificial Intelligence (AI) adoption in Compliance brings about many benefits and challenges at the same time. AI increases efficiency, accuracy, scalability and reduces the number of human errors in terms of automating complex processes, extracting patterns from large datasets and more. Real time monitoring of transactions, ability to assess risks and ensure compliance with evolving regulations can be done through AI driven compliance systems, thereby making an organization capable of maintaining compliance. There have been that some of these challenges, however, data privacy concern, algorithmic bias, and model transparency must be addressed for ethical and fair AI implementation. Implementing AI within compliance schemes will also require significant amounts of investment of the latest edge technology and human resources and an ongoing refresh in view of the evolving regulatory and legal landscapes. Organizations can fully leverage AI's potential as well as mitigate risks only in cases where it establishes strong governance frameworks, collaborates between AI developers as well as compliance professionals, and does a priority to ethical AI principles. Since AI is constantly developing, its involvement in compliance will become more important and thus, becoming more and more necessary to continuously improve and implement it responsibly.

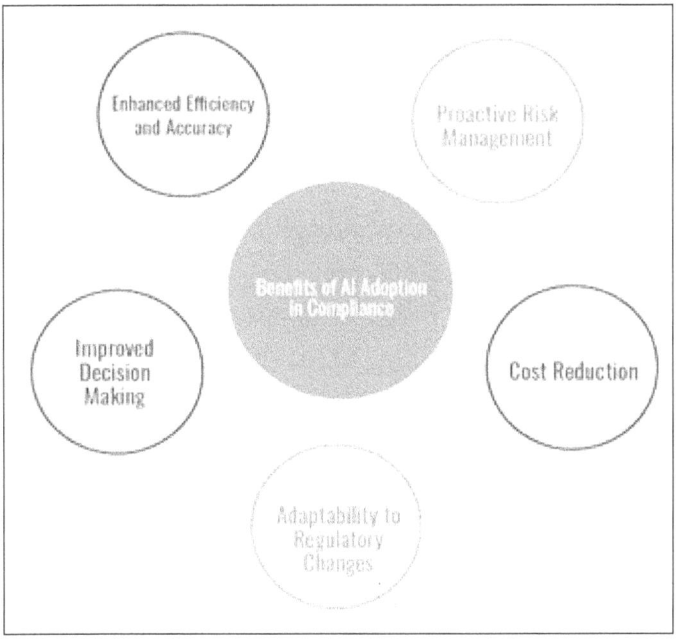

Figure 3.4 *Benefits of AI Adoption in compliance.*

Source: ***(Amblard-Ladurantie, 2024)***

- **Enhanced Efficiency and Accuracy:** AI significantly speeds up regulatory data processing and analysis, a core component of compliance tasks. This rapid processing and AI's ability to minimize human error leads to more accurate compliance reporting and decision-making.
- **Proactive Risk Management:** AI's predictive analytics capabilities allow organizations to anticipate potential compliance risks before they materialize. By analyzing trends and patterns in data, AI can identify areas of concern, enabling proactive rather than reactive risk management.
- **Cost Reduction:** AI can lead to considerable cost savings in compliance operations over time. By automating routine tasks, reducing the need for manual oversight, and improving accuracy, organizations can reduce operational costs related to compliance.

- **Adaptability to Regulatory Changes:** AI systems can be programmed to stay updated with compliance regulations and standards. This adaptability ensures that organizations can quickly adjust to regulatory changes, maintaining compliance at all times.
- **Improved Decision-Making:** With AI's comprehensive data analysis, compliance officers can make more informed decisions. AI provides deep insights that might not be evident through manual analysis, leading to better compliance strategies.

Challenges of AI Adoption in Compliance

Adoption of AI in compliance poses a set of challenges that organizations need to tackle to make AI in compliance effective, ethical, and regulation compliant.

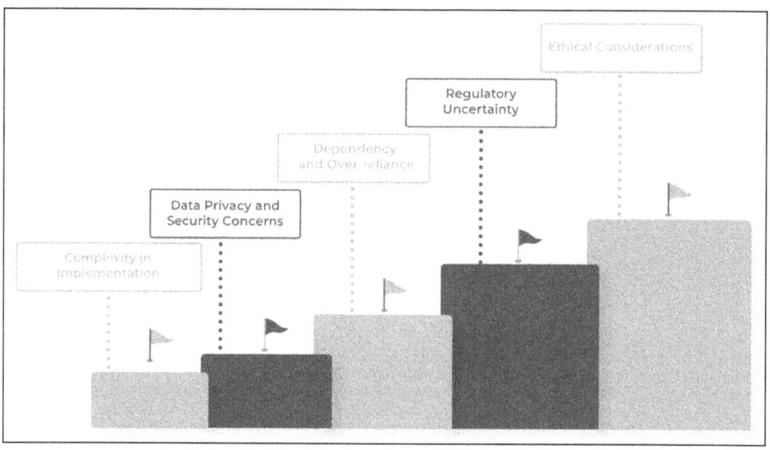

Figure 3.5 *Challenges of AI adoption in Compliance.*

Source: (Amblard-Ladurantie, 2024)

- **Complexity in Implementation:** Integrating AI into compliance frameworks can be complex and resource-intensive. It requires technical expertise, substantial initial investment, and an understanding of how AI can be best utilized within specific compliance contexts.

- **Data Privacy and Security Concerns:** AI systems rely on vast amounts of data, which raises concerns about data privacy and security. Ensuring that AI complies with data protection regulations is a significant challenge.
- **Dependency and Over-reliance:** There's a risk of becoming overly reliant on AI for compliance decisions. Maintaining a balance between automated and human decision-making is crucial to ensure that AI complements rather than replaces human judgment.
- **Regulatory Uncertainty:** AI in compliance is a relatively new field, and regulatory frameworks addressing AI's role in compliance are still evolving. Navigating this uncertain regulatory landscape can be challenging for organizations.
- **Ethical Considerations:** AI in compliance must adhere to ethical standards, particularly in decisions that might affect customers or employees. Ensuring that AI systems are fair, transparent, and non-discriminatory is crucial.

One of the challenges when adopting AI in compliance is a complex implementation, concerns over data privacy, over-reliance on the automation, regulatory uncertainty, and ethical considerations. To make an AI compliance stick, the organizations must address these problems in order to.

3.3.1 AI for real-time regulatory tracking and reporting

The regulatory environments are becoming more complex and fast changing and the businesses are facing greater challenges in complying with the industry laws and government regulations. Manual monitoring and reporting in these traditional methods are very time consuming and error prone. With AI based solutions providing a dynamic approach of real time regulatory tracking and automated report on compliance, organizations could control risks as well as inefficiency while still getting ahead of changes in laws and compliances. Artificial intelligence provides businesses with the ability to continuously ensure compliance with minimal manual intervention, making the accuracy and efficiency of its operations better (Robert Baldwin, Martin Cave, 2011).

Machine learning and natural language processing (NLP) constitute the science of AI systems which constantly monitor huge bodies of regulatory databases, legal and other documents, government updates and industry guidelines. They can analyze and understand the new regulations and figure out what it will mean to the business operation, and notify the compliance team about the things that need to be done. Unlike traditional manual methods, the compliance tracking with AI does not only decrease a business risk of legal issues and noncompliance penalties, but is done in an instant way, being immediately informed of regulatory changes. Additionally, AI categorizes regulatory updates by relevance so that businesses can better understand the importance and start implementing necessary adjustments to that end very quickly.

AI is also used to automate compliance reporting beyond tracking. Traditionally, model data collection, validation and interpretation must be performed through manual labor in multiple and large amounts, which can be a time and resource expensive and prone to errors. Automated by AI compliance systems which gather the data from various sources and verify it for accuracy before producing structured reports of the data. However, businesses can customize these reports to meet their many needs in regards to compliance documentation that can be submitted in a timely and error free manner. Similarity to this, AI driven reporting tools also assist in the audit preparedness by keeping track of the details and highlighting the differences that may occur before submission.

The other thing that AI does is predict risks, so it helps identifying risks that are not yet breaking the rules. AI systems will be able to predict potential areas of non-compliance using historical compliance data and patterns and recommend Wayne to take in advance. A proactive approach such as this allows businesses to prevent and eventually minimize risks and develop more effective compliance strategies. As AI technologies, machine learning, as well as deep learning become more and more adopted, it will always contribute to regulatory tracking and reporting, therefore providing more transparency, accuracy and agility in the compliance management.

Steps for Implementing AI in Regulatory Reporting

Integration of AI in regulatory reporting, though, has huge potential to completely disrupt the compliance processes, but with better planning and careful thought. In order to implement artificial intelligence in a way that will be effective for regulatory reporting strategies, organizations must follow a set of well-defined steps and best practices. Adhering to these regulations will help businesses improve the accuracy, a smooth reporting workflow, and comply with the rapidly changing regulatory practices.

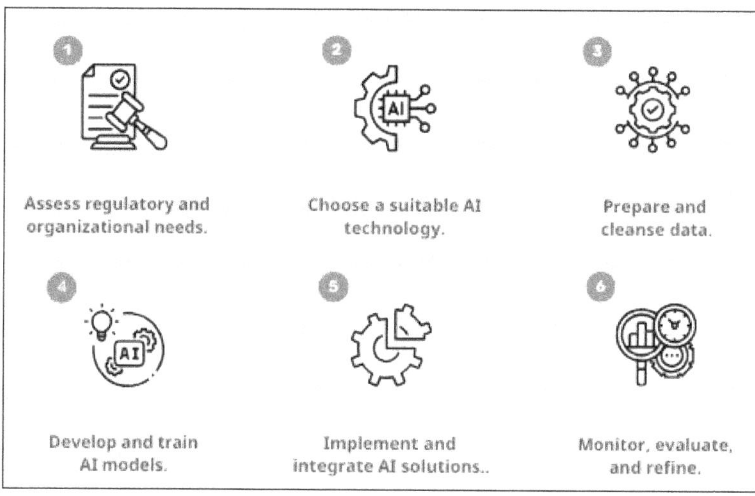

Figure 3.6 *Implementing AI in Regulatory Reporting.*

Source: (Meris, 2025)

1. Assess regulatory and organizational needs

In order to know where AI can help in regulatory reporting without compromising on business interests, it is important to know the organizational compliance requirements. A comprehensive compliance gap analysis will help organizations identify efficiency and weakness of existing reporting workflows; this, in turn, will help organizations develop productive AI driven approaches to improve their accuracy and automation of reporting procedures. To be implemented with proper regard, engagement of compliance officers, IT teams and legal experts is important to obtain proper alignment of AI with regulatory requirements and industry practices. In addition, organizations have the ability to reap early benefits by concentrating on high impact areas, for example, where data is intensive, or where reporting is frequent. This improves efficiency and reduces manual errors. This focus on these critical aspects can be leveraged to lay the solid foundation for the wider adoption of AI in regulatory compliance, for achieving higher accuracy, minimizing risks, and speed up the operations.

2. Choose a suitable AI technology

AI focused compliance solution must be scalable, adaptable, and integrable all the time. The solution chosen must address current regulation as well as anticipate the change in regulation. This motivates platforms with easy to use interfaces and extensive support systems as they foster adoption among the employees, which in turn results in smoother transitions and maximum operational benefits. Knowing that you will rely on a solution that has experience in regulatory reporting gives you peace of mind when leveraging it. Furthermore, integrating well with existing enterprise systems, for example, Enterprise Resource Planning (ERP) preserves the workflow continuity and data consistency. It is also important to assess the provider's data security measures and check the compliance with the relevant legal frameworks, such as the General Data Protection Regulation (GDPR), as they also act as a measure to protect sensitive information as well as regulatory compliance.

3. Prepare and cleanse data

Before deploying AI systems, organizations should standardize and cleanse their data and organize it for increased accuracy and reliability of regulatory compliance processes. Data clean and structured is a guarantee that the AI models will produce sense giving insights and not end up in error or inconsistency. Likewise, a well-prepared data foundation also increases scalability and flexibility as new datasets are introduced making it easy to integrate with new as well as emerging compliance requirements. For a better data governance, we need to implement robust data governance frameworks so that data integrity is intact and the same practices are practiced throughout the organization. Using automated data cleansing tools helps to clean up the data inconsistencies, redundancies or inaccuracies which would improve the overall AI driven decision making in the network. Further, this enables easier processing of the data and accelerates training of the AI model, which in turn reduces time for compliance reporting.

4. Develop and train AI models

In order to reflect on the regulatory environments and reporting requirements, organizations should train their AI models with appropriate data. Training of an AI system is necessary for it to understand the industry specific compliance nuances and thus its ability to detect anomalies, identify risks, and generate an accurate report. Organization can render AI models smarter using the historical compliance data and reports not only to notice patterns but also to make predictive accuracy. During the training phase, if subject matter experts are involved, there is a lower risk of compliance errors as AI outputs are validated for accuracy and relevance to the regulatory requirements. Also, a small-scale rigorous testing will help companies discover issues that may occur before they really go to full deployment. An approach toward AI training that is structured helps to increase the reliability and effectiveness of pre-built compliance automation so that your regulatory adherence and your risk management are enhanced.

5. Implement and integrate AI solutions

The structured rollout of AI solutions is the way forward for organizations as it builds capacity for ready integration with the existing systems like accounting software and compliance management platforms. Technical integration with user adoption in a balanced way is needed to improve operational efficiency without disturbing the workflow. To make the most of benefits from the AI system, it is important to provide comprehensive employee training so that staff can use the AI system effectively. Phase implementation is also an approach that eases the operational disruption with ability to go online and tweak as we go live based on user feedback. Moreover, setting up monitoring protocols with clearly defined Key Performance Indicator (KPIs) also allows organizations to continuously evaluate the AI performance, measure success, and modify the system accordingly. A well-planned rollout will improve adoption rate, improves compliance processes and strengthens enforcement.

6. Monitor, evaluate, and refine

AI systems need to be continuously monitored to keep them on track with changes of the regulatory requirements and business objectives. That allows businesses to periodically assess incomings risks, perfect with AI, and improve the efficiency of compliance over time. By updating AI models frequently, they no longer violate the most recent laws or the trends in the industry. It is important to seek feedback from the users and all the stakeholders so as to know which areas are related to improvement, how to refine the system usability and the overall efficiency of the system. Periodic audits also serve the important purpose of ensuring the accuracy of AI generated reports as well as to boost trust and compliance integrity with regulators. An approach to proactive monitoring will guarantee continuous effectiveness, adaptability, and ability to meet the requirements of the regulatory compliance.

3.3.2 NLP applications in compliance automation

Automation of NLP enable us to analyze regulatory texts, policies and communications in order to understand compliance monitoring. In finance, healthcare and energy, where industry and regulations are often complex, updated and mandatory, it is essential for an organisation to obey legal and industry standards strictly. NLP can power systems to process massive amounts of unstructured data, extract the key regulatory clauses and compare them to the policies of the company to find the discrepancies. It allows the organizations to remain up to date with the compliance requirements while easing the burden of manual work on legal and compliance teams. NLP summarizes regulations and highlights to its users the most recent compliance hot parts, so that when the landscape is evolving, organizations can change immediately.

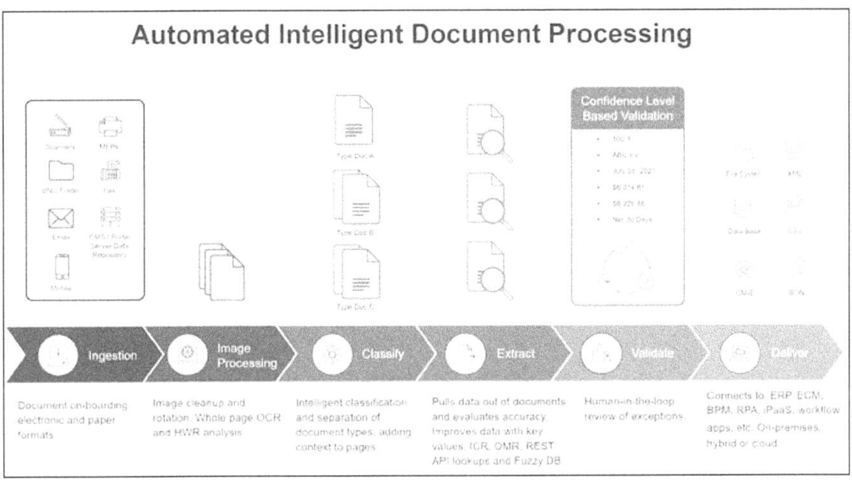

Figure 3.7 *Automated Intelligent Document Processing.*

Source: (Ahmed, 2023)

In addition to that, NLP enables real time monitoring of corporate communications to, for example, detect potential compliance violations. Emails, chat messages and phone transcripts are closely monitored by many industries due to fraud, insider trading and data breaches. NLP based solutions help analyzing language patterns, flag suspicious activities and alerting compliance officer to the potential risks. Furthermore, these systems reduce the auditing work by automatically scanning and sorting out compliance related documents to detect anomalies and generate accurate reports. With such a proactive manner of monitoring, legal risks are significantly reduced and organization are guaranteed to hold minimal disruption in abiding by laws.

Additionally, NLP aids in predictive compliance analytics where businesses can predict and prevent regulatory risks from rising. NLP models review historical data and observe trends to make a prediction of potential compliance problems and suggest possible solutions. It not only strengthens risk management strategies, but also enables organizations to make decisions without getting penalized handsomely and offering up their reputation. With increasing regulatory requirements becoming more complex, organizations will rely more and more on NLP based solutions to increase regulatory oversight but also to maximize operational efficiency.

3.4 AI in Crisis Management and Business Continuity Planning

Today most affairs take place in a few seconds in the fast economy, thus there are many crises that businesses should deal with including natural disasters, cyber attacks, economic freezes and supply chain stall. Effective crisis management is crucial for corporate resilience and helps in managing the business continuity planning. Artificial Intelligence (AI) has emerged as a very powerful tool to predict the risk, make the real time decisions and auto response in this domain. However, AI enabled analytics enable those businesses to anticipate the vulnerability, anticipate potential crisis and data driven strategies to prevent disruption when it happens (Kurt J. Engemann, 2014).

With regards to crisis management, AI is one of the key applications of AI as they include real time monitoring and early warning systems. The platforms use AI to analyse huge volumes of structured and unstructured data offered from news reports, social media, IoT sensors and the internal business systems to identify anomalies and predict possible risk. For instance, machine learning models can be used to indicate a cybersecurity threats, the extreme weather conditions and the supply chain bottlenecks, so that the businesses could act in advance. Chatbots and virtual assistants employing AI assist a great deal in crisis communication by flooding employees and customers with the latest updates and guiding them with it, and ensuring that it reaches every corner in as little time as possible.

Apart from saving time on crisis detection, AI saves corporate spending by automating crisis recovery processes and resource allocation. Such systems aid organizations in building dynamic contingency plans through simulation of various crisis scenarios and recommendation of the optimal response strategy. Say, in the case of an IT system failure, it might automatically trigger to start up the automatic disaster recovery protocols, re-route network traffic and time critical data from backups. Just like this, the supply chain optimization tool provided by AI will be able to establish alternative suppliers and redirect the logistics operations to minimize the disruptions. For this reason, AI will be increasingly important to use in the purpose of crisis management and business continuity planning to help business operations continue as normal when the future of business is uncertain.

3.4.1 AI for disaster response and risk simulation

The use of Artificial Intelligence (AI) in disaster response and risk simulation speeds up, improves and data orientated decisions in crisis situations. Quicker intervention is needed to cover humanitarian assistance in the case of natural disasters like earthquakes, Hurricane, wildfires, pandemic and floods which require immediate and well-coordinated intervention to those casualties, protect the critical life line important in development, and reduce the economic losses. Aided by AI, the disaster preparedness comes with predictions of possible dangers through leveraging AI, optimization of emergency response efforts and automation of resource allocation. Machine learning techniques that can leverage the use of machine learning models trained on datasets containing large amounts of historical disaster records, real time satellite imagery, environment IoT enabled sensors and as well as meteorological data can forecast disasters with much higher accuracy. As a result, governments, humanitarian organizations, and business are empowered to proactively think of ways to minimize the risks of a crisis even before it begins (Usha, 2024).

AI is very useful in disaster response as it is of help improving situational awareness, coordination and efficiency. These same intelligent drones and autonomous robots assist in search and rescue mission by conducting missions under hazardous and inaccessible environment, locating survivor trapped under debris, determining structural damage of affected area and mapping in a precise manner. Advanced Natural Language Processing (NLP) algorithms mine large quantities of emergency call data, social media updates, satellite feeds, as well as news reports enables them to detect distress signals, determine where emergency resources should be focused, and allowing first responders to keep up with the live intelligence. AI combined with raw open data from several sources delivers powers to emergency management decision support systems to help emergency management team to plan, to prioritize resource distribution, to design optimal evacuation route, and alike to deploy medical aid, food and emergency shelter as effectively as possible. You know, by doing auto, critical thing in disaster response, AI significantly reduces disaster response time, improve logistically coordination and save lives and mitigate the ground effect that will come in future disasters.

Figure 3.8 *AI-Powered Crisis Response and Disaster Management.*

Source: (Artmaster, 2024)

Beyond its effective use in immediate disaster response, AI based risk simulation enhances preparedness by simulating different crisis scenarios in a way that allows their impact be evaluated in various conditions. This is a simulation based on what they (AI) are and how they make decisions, geospatial data, urban infrastructure vulnerabilities and socio-economic factors, and they predict what disaster will come and what the best mitigation strategies will be. By incorporating these advanced modeling techniques, urban planners, urban policymakers, emergency responders and municipalities can design resilient infrastructures, optimal land use planning, a data driven disaster planning using their data and strengthening community resilience. Digital twins, powered by AI, function by creating virtually real replicas of physical areas that enable government fiscals and businesses to experiment with various disaster response strategies to make them better prepared and better to face future risks. While an advanced level of AI is already at work in the disaster response and risk simulation, it will continue to grow and will apply to new technologies such as edge computing, blockchain for transparent relief fund distribution, and 5G-enabled smart sensor networks to add further sophistication to its role in these tasks for more safer, adaptive and resilient world.

3.4.2 AI-driven strategies for proactive risk mitigation

Artificial Intelligence brings a new twist to risk mitigation which enables businesses, governments and the organizations to anticipate and prevent the emergence of any threat to turn into crisis. Such AI based risk management approaches are extremely different from the usual reactive approaches, as it involves using predictive analytics, machine learning models and real time processing of data to run vulnerability and spot vulnerabilities before they occur. Once industries have used it, industries become more resilient, make better decisions and avoid financial, operational and reputational risks by building on the insight given by AI (Skene, 2019).

For instance, in the field of cybersecurity risk mitigation, AI safety, financial risk mitigation, supply chain resilience, infrastructure safety to climate risk assessment are based on risk mitigation strategies based on AI. Advanced data analytics means that the anomalies can be detected with AI; failures can be predicted and the system proactively intervened to avoid disruption. Over time, these organizations have become so dependent on the AI powered automation to add muscle to their risk management framework through prime monitoring and adaptive response to spotted threats. The other aspect is regarding the integration of AI with IoT and blockchain and cloud computing which further mobilizes its potential to detect and mitigate the risk comparatively seamlessly across the global networks.

As AI continues to develop, it is expected that AI will control an increasingly greater component of the proactive risk mitigation and there will be increasingly greater accuracy and efficiency in spotting potential threats. In industry, AI is used for predictive maintenance, in the finance for detecting fraud, and in any sector that has to take care of risk for early intervention and prevention. Digital twins and AI autonomous response mechanisms empowered by the simulation models of AI make possible the organization's simulation of hypothetical risk scenarios and optimization of mitigation strategies. The use of AI led risk mitigation frameworks allows businesses and governments to bypass the current uncertainties and reach a long duration sustainability and security.

3.5 Industry-Specific Risk Mitigation Strategies

There are different risk mitigation strategies across industries and the approaches that are required will be specific to the industries facing different problems. Today, in the fast-changing global market, businesses should be actively looking for risks, assessing them, and minimizing them to ensure the stability of their operations and their financial standing and reputation. For industries to be stable and be a success in the long term, they must anticipate, assess and respond to risks very precisely. In banking, industry specific risk mitigation strategies are important to ensuring that they remain free of financial fraud, as well as in healthcare, so that patient safety is maintained, as well as in manufacturing, so that businesses can continue to operate efficiently (Spedding & Rose, 2018).

With the help of such advanced technologies, the regulatory framework, and data driven decision making they can predict the potential risk and undertake counter measures ahead of time when they occur. Industry specific strategies for financial fraud prevention, predictive maintenance in manufacturing, decision making, resilience, and crisis response are all increased through the use of these strategies. This chapter looks into the different industries and what key risk mitigation strategies should be to some of today's industries, best practices, and innovations that help organize to be considerations of potential threats and uncertainties.

Industry-Specific Risk Mitigation Strategies are:

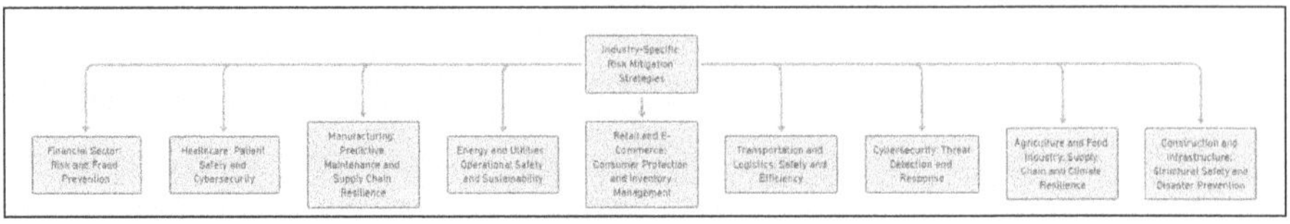

Figure 3.9 *Industry specific risk mitigation strategies.*

Source: (Self-Generated)

1. Financial Sector: Risk and Fraud Prevention

Risk includes fraud, credit default, market volatility and compliance breach faced by the financial industry. Transaction monitoring and behavioral analysis are the basis for the detection systems of fraud. Credit risk assessment models are applied by banks and insurance firms to determine the financial health of borrowers thus reducing the possibility of default. With regulatory compliance frameworks, institutions ensure ethical and financial practices through stabilization of the market. AML policies are quite strong to make sure the financial institutions are up to date with global regulations and avoid any type of financial activities.

2. Healthcare: Patient Safety and Cybersecurity

The risks to healthcare industry include such things as medical error, data breaches, and drug safety issues. Structuring of a decision flow or a structured workflow helps to reduce misdiagnosis rates by enabling accurate analysis of the medical data. These risk management strategies of pharmaceutical are to identify adverse drug interactions prior to clinical trials. Ensuring security in Drs's digital world is still a top priority to Cyber threats but robust security measures are in place to safeguard Electronic Health Records (EHRs). Compliance with strict regulations helps minimize the chances of breaches of data protection and helps in maintaining confidentiality in patient data.

3. Manufacturing: Predictive Maintenance and Supply Chain Resilience

Operational risks faced by the manufacturers are equipment failures, supply chain disruptions and quality control issues. Regular equipment inspections and condition-based monitoring forms basis of predictive maintenance programs to prevent breakdowns that are very costly. Risk assessment methods that warn about raw materials shortages and optimums procurement methods improve the supply chain resilience. Walvoord states that digital modeling and process simulations can help manufacturers improve operational efficiency while ensuring that a product remains of quality. Quality control measures are strictly applied to eliminate defect products before they are given to the consumers.

4. Energy and Utilities: Operational Safety and Sustainability

Since the energy sector is fraught with threats like grid failures, environmental hazards, and uncertain energy demands, it is hard to make wise decisions and to attain a maximum return on investment (ROI). The power grids are stabilized using predictive analytics and monitoring systems so as to observe historical and real time data to avoid outages. In the oil and gas industry, safety includes continuous monitoring of the pipelines and equipment to detect any leak and malfunctions. Resource management is applied by renewable energy providers to ensure the balance between supply and demand. It helps the energy companies to minimize carbon footprint and enhance their sustainability initiatives.

5. Retail and E-Commerce: Consumer Protection and Inventory Management

Fraud, data breach, inventory mismanagement risks need to be mitigated for retailers and e-commerce platforms. Real time fraud prevention systems pick out unauthorized transactions and fraud activities. By invoking the demand forccasting tools, the retailers are able to ensure optimal inventory levels by reducing stock out, and over stocking. Measures for security like biometric authentication and transaction monitoring are taken to protect on line consumer from frauds. Logistics planning and supply chain risk assessments are conducted properly so as to dispatch the product in time and be able to manage operational disruptions.

6. Transportation and Logistics: Safety and Efficiency

There are risks like accidents, route inefficiencies, logistical bottlenecks and the transportation industry has to address them. Traffic management strategies aim at identifying accident prone areas and improving air route to reduce the risk of road accidents. Advanced fleet management system assures regular vehicle to be maintained, which in turn reduces breakdowns and delays. Risk assessment plays an important part in reducing supply chain disruptions during logistics operations. Transportation companies implement comprehensive safety methods to seek the protection of assets and also how they ensure keeping the industry's requirements.

7. Cybersecurity: Threat Detection and Response

The risks of cyber threats are very high like data breaches, ransomware attacks, and phishing scams. Network traffic is analyzed by security monitoring systems to find information which might indicate a breach has occurred or is about to occur. Encryption and access control policies are strong that unauthorized data can be accessed. Cyber incident response plans expedite containment of the threat to the minimum possible extent. Employee training programs create awareness of cybersecurity risks and reduce the chance of internal vulnerability.

8. Agriculture and Food Industry: Supply Chain and Climate Resilience

Risks such as climate change, crop diseases, food contamination, affect agriculture and food supply chains. These crop monitoring systems aid in tracking environmental conditions to forecast droughts, pest infestations, bad soil health among others. Monitoring temperature sensitive products and risk of contamination is one-way supply chain tracking mechanisms help provide food safety. Climate risk assessment helps the farmers to adapt to the changing weather regimes by efficiently managing irrigation and resource allocation. The application of sustainable farming practice aids to improve yield projections and ensures food security.

9. Construction and Infrastructure: Structural Safety and Disaster Prevention

Risks such as structural failures, project delays, worker safety hazards present threats to the construction industry. Suspension bridge is monitored by this sensor-based system which use structural health monitoring systems to give an assessment of early signs of wear of such buildings, bridges and dams. Compliance with occupational health regulations in construction sites reduces accidents and also enforces safety protocols in construction sites. Digital modeling techniques enable engineers to create simulations during which the execution of infrastructure projects can be assessed, their risks can be evaluated, and their construction plans can be optimized, before ever executing on it. The disaster preparedness frameworks assist in estimating the impact of natural calamities on the infrastructure and formulating mitigative strategies.

3.5.1 AI applications in healthcare, finance, and manufacturing risk management

Artificial Intelligence (AI) has changed the face of Risk Management in several ways across multiple industries by improving the predictive power, streamlining the immense processes and assuring better decisions. AI driven solutions mitigate business risks in business sectors like finance, healthcare, and manufacturing that pertains to compliance, operational failures, cyber security threats or financial losses. AI uses machine learning, natural language processing and real-time analytics in order to beneficially predict what can happen and to prepare industries proactively from the risks, minimize the harm of disruptions and improve overall resistance. Below are explored in the following sections how AI is affecting risk management in these key industries (Das, 2024).

1. AI in Healthcare Risk Management

There are many risks for the healthcare sector such as misdiagnosis, cybersecurity threat, operational inefficiency and regulatory compliance issues. AI powered risk management solution provides a safe environment for patients, provides better hospital operations and better security of data.

- **Predictive Diagnostics and Patient Safety** –AI models apply predictive diagnostics and patient safety by analyzing huge data of patients' pasts to detect the early signs of illnesses, which makes misdiagnosis rates decline and the accuracy of treatment sky rock steadily. The machine learning algorithms help doctors in identifying potential problems associated with the treatment and suggesting personalized treatment plans.
- **Cybersecurity in Electronic Health Records (EHRs)** – AI helps to secure the data through the identification of suspicious activities in real time and preventing a person from accessing the patient's sensitive information. Data breaches are reduced with the use of advanced encryption technique and AI involved authentication systems.
- **Operational Efficiency and Workflow Automation** – AI automates tasks like patient schedule, coding of medical and claims processing that eliminate human errors and how well the resources are allocated.
- **Regulatory Compliance and Risk Monitoring** – AI has the ability to help healthcare organizations to track and stick

to the changing regulatory requirements and adhere to the industry standard such as HIPAA and GDPR.

2. AI in Financial Risk Management

However, AI can be used by financial institutions like banks to detect fraudulent transactions, credit risks assessment, implementation of investment strategies, and meeting regulatory compliance. The models for use of AI are driven by the fact of being real time with insights to mitigate the loss in financial and to enhance the decision making.

- **Fraud Detection and Prevention** – A fraud detection and prevention AI based system that employs AI based fraud detection system to analyse transactional data and detect anomalies as well as suspicious activities in real time. The machine learning models keep retraining itself to the new evolving fraud pattern, thus reducing the false positives and increasing the accuracy.
- **AI for Credit Risk Assessment** – It assesses credit risk by profiling borrower's historical financial data, transaction patterns as well as by other data sources to arrive at its ratings. This allows them to make informed lending decisions for banks and lending institutions.
- **Market Risk Prediction**–With financial trends, market and economic indicators as references, the AI algorithms try to predict risks on investments. Robo-advisors made with the help of AI, provide the investors with help to make data driven decisions.
- **Regulatory Compliance and AML (Anti-Money Laundering) Monitoring** – Financial crime is automated with the help of AI for regulatory compliance and AML (Anti Money Laundering) monitoring that checks the regulatory requirements, suspicious transactions, and follows appropriate financial regulations. For reviewing contracts and financial statements for compliance violations, there exist NLP tools that help with that.

3. AI in Manufacturing Risk Management

Risks associated with equipment failure, supply chain disruption, quality control and worker safety are some of the risks in the manufacturing industry. Utilizing artificial intelligence (AI) powered solutions, manufacturers boost productivity, anticipate possible failures, as well as resort to operational resilience.

- **Predictive Maintenance and Equipment Reliability** – predictive maintenance systems powered by AI are trained to look for early signs of machine failures in order to reduce downtime and reduce expense under maintenance. Manufacturers can thereby schedule maintenance before component failures.
- **Supply Chain Risk Mitigation** – AI improves the supply chain visibility by monitoring logistics, detecting the possibility of threat and optimizing inventory management. Demand forecasting models that use AI keep the manufacturer on guard for material shortages and change the production schedules.
- **Quality Control and Defect Detection** – Computer vision powered systems are used for quality control in real time and drastically reduces the production waste. It reduces the human errors and increases efficiency.
- **Workplace Safety and Risk Assessment** – AI improves workplace safety through analysis of the environmental conditions, monitoring developments in compliance with safety regulations and prediction of possible risks. Wearable devices of AI wearables monitor workers' health and alert supervisors in case of emergencies.

3.5.2 AI's role in enhancing organizational resilience

Disruptions, challenges, and crises have become an uncertain reality for organizations in the world that we live in. Our use of

Artificial Intelligence (AI) can help it out in enhancing organizational resilience—they can give real time data analysis, predictive insights and might be able to do the automated decision making. AI driven risk assessment models provide companies a view of the possible threats (cyber-attack, supply chain disruption, financial fraud or others) so they can take action based on the identification of potential problems before they reach crisis levels. Organizations can keep monitoring the market trends, regulatory changes and operational risks with the help of machine learning algorithms and natural language processing, thereby making the organization continue to be agile and reactive to the challenges emerging in the market (Ummah, 2019).

AI helps increase resilience by making the use of resources and operation more efficient. AI automation using automation is now being adopted in the industries like healthcare, manufacturing and finance, where it eventually reduces human errors and improves workflow with productivity. For instance, predictive maintenance systems at manufacturing spots detect initial warning signals of equipment's collapse a plenty of time before maintenance turnaround can lead to expense downtime, whereas AI powered security solutions spy and pinpoint cyber risks to critical infrastructure before they have the opportunity to infect it. Furthermore, the use of AI helps to make employees more resilient in the workforce by offering intelligent automation solutions that help employees make strategic decisions rather than performing repetitive tasks, which makes a company more flexible according to dynamic circumstances.

Resilience beyond the risks is being enabled by AI, as it is capable of continuous learning and innovation. AI analyses data at a historic and real time base, improving organizations' strategies, enhancing the mechanisms of response to the crisis and the development of more efficient contingency plans. Digital twins powered by AI enable businesses to simulate different types of risk scenarios, instead of reacting without any plans in the case of the occurrence of unforeseen events. Additionally, companies use AI for customer insights to acquire an understanding of how their consumers are changing in this rapidly evolving market, which keeps them ahead of the game. With AI technology getting better, the additional importance of its contribution to organizational resilience will become more prominent, allowing businesses to have navigating of uncertainty and sustain themselves through.

Multiple Choice Questions (MCQs)

1. Which of the following is a key benefit of AI in risk management?

A. Increasing manual paperwork

B. Reducing predictive capabilities

C. Enhancing risk identification and mitigation

D. Limiting data-driven decision-making

2. How do AI models assist in predictive risk assessment?

A. By eliminating the need for human intervention

B. By analyzing historical and real-time data to anticipate risks

C. By manually reviewing compliance reports

D. By ignoring external market factors

3. Which AI technique is commonly used for operational risk analysis?

A. Rule-based automation

B. Machine learning models

C. Handwritten risk reports

D. Manual forecasting

4. What is a key role of AI in compliance monitoring?

A. Generating regulatory frameworks manually

B. Tracking regulatory updates in real time

C. Replacing legal professionals

D. Ignoring data privacy laws

5. How does NLP contribute to compliance automation?

A. By automating document review and regulatory interpretation

B. By replacing financial audits

C. By generating synthetic legal cases

D. By removing compliance requirements

6. What is one-way AI supports crisis management?

A. By delaying disaster response strategies

B. By using AI-driven simulations for risk preparedness

C. By eliminating emergency protocols

D. By focusing only on past crisis events

7. What is a benefit of AI-driven strategies for proactive risk mitigation?

A. Reducing the need for strategic planning

B. Enhancing decision-making by forecasting risks in advance

C. Ignoring supply chain risks

D. Avoiding regulatory oversight

8. How does AI assist in industry-specific risk management?

A. By providing tailored risk analysis for sectors like healthcare, finance, and manufacturing

B. By standardizing risk management across all industries

C. By removing the need for human oversight

D. By ignoring sector-specific regulations

9. In what way does AI enhance organizational resilience?

A. By automating crisis response and operational efficiency

B. By eliminating the need for cybersecurity measures

C. By disregarding external risk factors

D. By replacing human employees entirely

10. What role does AI play in financial risk management?

A. Detecting fraud and automating compliance monitoring

B. Eliminating credit scoring models

C. Increasing manual transaction analysis

D. Ignoring financial regulations

Answers

1	2	3	4	5	6	7	8	9	10
C	B	B	B	A	B	B	A	A	A

Chapter 04

AI in Financial Security and Fraud Detection

4.1 Introduction to AI in Financial Security

Artificial intelligence (AI) in finance refers to the application of a set of technologies, particularly machine learning algorithms, in the finance industry. This fintech helps financial services organizations speed, reduce and improve the accuracy of such tasks as data analytics, forecasting, investment management, risk management, fraud detection, customer service, and more. Currently, AI is modernizing the financial industry by automating traditionally manual banking processes, as well as providing better insights of financial markets and enabling ways to interact with customers in ways that are human intelligent and human like.

Financial institutions are being revolutionized by AI, and star ups are being fuelled. With the aid of real time market data, AI models execute trades with what would have to be considered almost unimaginable speed and precision and dictate where investments will be based. AI solutions help financials to optimize on those complex patterns found in transaction data sets for risk management work including security, fraud, AML, KYC, and more. Similarly, AI is also redefining how financial organizations interact with customers and understand their behavior as well as purchasing preference. This allows for more personalized and faster customer support, credit score refinements, innovative products and services.

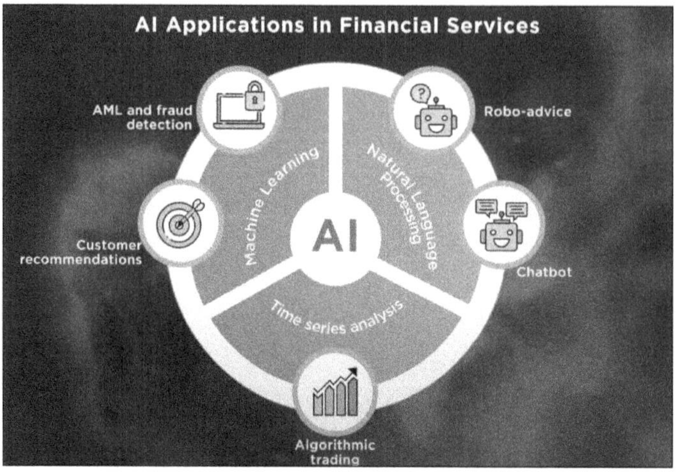

Figure 4.1 *AI Application in Financial Services.*

Source: (Maxim Dupont, 2023)

Main Components of AI in the Image:

1. Machine Learning (ML):

- AI systems which are not programmed explicitly are enabled to learn from data and to get better by machine learning.
- It is used to play a very crucial role in financial applications, fraud detection and customer analytics.

2. Natural Language Processing (NLP):

- NL helps AI to understand, interpret and generate human language.
- Since it is required in applications such as chatbots, virtual assistant and robo advisory services.

3. Time Series Analysis:

- This technique deals with analyzing data points collected at different points in time.
- These are all widely used in financial forecasting, algorithmic trading, predictive maintenance, etc.

Elaboration of AI Applications

- **AML and Fraud Detection:**

Identifying suspicious or even suspicious patterns of transactions that may be indicative of illicit financial activities is important in addition to AML and fraud detection with the help of Artificial Intelligence. They work as machine learning algorithms that analyze the huge amount of transaction data for finding anomalies, unusual behavior or pattern deviations from normal. The fraud detection systems based on these AI technologies will real time alert about potentially fraudulent activities thereby reduce financial losses and also ensures regulatory compliance. AI is also used by banks and financial institutions to prevent fraudulent transactions from ever reaching customers by increasing security.

- **Customer Recommendations:**

A key element in improving user experience through AI that helps turn customers into rate and review bases is an AI recommendation system that analyzes consumer behavior, purchase history, and preferences. As it relates to AI, E-commerce platforms such as Amazon and streaming platforms such as Netflix use AI to select the products tagged as relevant to certain consumer and view their specific interactions with them. These are machine learning based systems that use algorithms to predict the user preferences and increase customer satisfaction along with marketing. As AI is able to learn and improve recommendations over time, the users will have more and more accurate and personalized suggestions.

- **Algorithmic Trading:**

Powerful AI algorithmic trading runs out through financial markets to quickly execute trades with speed. It helps refine the analysis of market trends, historical data with real time news to allow them to trade inform their decisions within milliseconds. Using time series analysis, these systems give us the ability to predict fluctuations of stock price and trade in a way that is optimized. It automatically reduces human intervention, minimizes the risk, and maximizes profitability for the traders and investors. Overall, especially for hedge fund and financial institutions, they rely on the use of AI based trading systems to give them a competitive edge over other markets globally.

- **Chatbots:**

NLP powers the AI chatbots to interact with users, process the queries and automate customer support. Banking, healthcare, and retail are the sectors which include these chatbots to offer instant assistance to the customers without human intervention. Routine inquiries, common issues and even transactional support are handled in this manner, which reduces the customer service efficiency. Continuous learning means that the advanced chatbots keep evolving based on their interactions and are able to answer accurately and within the context.

- **Robo-Advisors:**

Robo advisors (robo advisors) based on AI transform the wealth management concept by providing automated financial planning as well as automated investment advice. These are financial analysis-based systems that analyse a user's financial goals, risk tolerance and market conditions and offer investment suggestions based on it. The usage of robo-advisors helps eliminate the requirement of human financial advisors, enabling those people gaining that the capacity to invest at less cost and with more handle. AI powered insights enable the investors to create a balanced portfolio so that you save some money and the best complement the strategies according to changing market conditions to be smarter and more efficient financial planning.

Key stakeholders of AI in finance

There is a wide variety of stakeholders in the implementation and operation of AI in the financial sector, each with a key role for the AI driven system to be effective, secure, and compliant. Awareness of the potential pitfalls of applying AI to a business' operational processes brings auditors and internal control teams to examine AI models for accuracy, efficiency, and regulatory adherence, as well as CIOs and CTOs who are responsible for overseeing the embedding of AI within an organization's technical infrastructure. It's the developers' job to design and make the AI algorithm as optimized as possible for reliability and fairness. Along the same lines, ethics and diversity officers try to curb biases and promote diversity in the use of AI. The decision to adopt AI is a function of business decisions made by executives and board members, they consider innovation with regulatory and ethics issues.

AI is widely utilized in banks and investment firms for fraud detection, management of risk, as well as to offer better customer service, and is hence considered as a key element of modern financial operations. Regulators and legal teams work with regulators to ensure that AI has been developed to comply with industry laws and the protection of customer data, while risk management teams monitor and monitor the models to check for any financial threats and avoid losses. While customers are the final users of AI driven financial services, they are essential in the success of AI adoption for their trust and engagement. Together, these stakeholders make sure that financial sector good uses of the AI technologies by making still efficient, still secure and still making better decision.

Use cases of AI in finance

AI is used by a variety of financial organizations to enhance user experience (UX), decision-making, and efficiency. Here are a few instances of AI in finance:

- **Customer service:** Chatbots powered by conversational AI and natural language processing (NLP) enable banking clients to obtain account information promptly and effectively, around-the-clock.

- **Preventing cyberattacks:** AI can employ data science to examine patterns and trends and notify businesses of anomalous activities.
- **Financial planning:** Based on clients' objectives, risk tolerance, and market conditions, robo-advisors employ complex algorithms to offer reasonably priced, individualized investment advice.
- **Fraud detection and prevention:** When anomalous spending patterns are found, deep learning can be used to examine consumer purchasing trends and send out an alarm.
- **Loan eligibility:** To reduce risk, lenders use AI neural networks to swiftly evaluate data and assess borrowers' creditworthiness.
- **Trading:** Algorithmic trading, or quick transactions based on current data and market trends, is how investment firms employ AI.

4.2 AI-Driven Transaction Monitoring and Fraud Prevention

Today, the financial landscape is entirely different from what it was mere years back due to the development of digital banking, online payment systems, as well as mobile financial services. This has resulted in an increase in volume, complex financial transactions, which is a difficult task to detect and prevent fraud with traditional accounts. Fraudsters are now outwitting old legacy systems that are usually using static rule-based approaches or human involvement and can no longer respond to the increasingly sophisticated tactics used by fraudsters. Unsurprisingly, the emergence of Artificial Intelligence (AI) as a powerful tool to strengthen fraud prevention measures in the financial sector while protecting the transactions is one of the latest cultural phenomena.

AI improves transaction monitoring to finance institutions with more intelligent and real time and scalable solutions. Traditional systems have the limitation to only detect the new or evolving fraud schemes as they depend on predefined rules based on historical data. However, in contrast, AI systems (powered by machine learning algorithms and big data analytics) are able to learn from a huge supply of transaction data and change with new evolving threats all the time. It facilitates fast, precise identification of suspicious activities and indicating possible fraud by AI.

Real-time analysis of the large volume of data that is processed by token monitoring engines is one of the major benefits of having AI-driven token monitoring engines. Considering the percentage of transactions that take place every second within today's digital banking ecosystem, human agents or rule-based systems would not be able to monitor each transaction individually. But AI can also parse through millions of data points in a split second to identify these anomalies that may consider buying and selling fraudulent activity. For instance, AI can set a flag on any such transaction, for example large withdrawal from an unknown address or a sudden change in spending patterns, and if this happens, the AI can immediately intervene with an automated response, such as accounts getting frozen or requesting additional user authentication.

Furthermore, the learning ability of AI from new data keeps accumulating and enables AI to find patterns of fraud that are difficult to find with traditional methods. If the models can't adhere to all of the patterns they are preprogrammed with, they are able to recognize subtle correlations in data that may signify fraud, and they even go so far as to take note of the fact that sometimes they don't do what they are expected to do. This enables AI systems to not only be reactive but also preemptive, as it sharpens their ability to learn from every new attempt to defraud an AI system.

The use of AI in fraud prevention is another critical advantage due to the fact that it blends different data sources to boost decision-making. An AI will look at a great many factors in order to gauge transaction risk (such as device data, transaction history, geolocation data, and standard user behavior patterns). From this holistic view, AI is better able to assess what constitutes a risk each time a transaction is made and if it is significant enough to warrant further investigation or immediate action.

4.2.1 Real-time fraud detection with AI

AI misuse has been made possible by its widespread accessibility and lack of a clear rulebook. The reputation of artificial intelligence has been damaged by bad use cases, and many people are unaware of the many positive uses of AI. AI algorithms, for instance, can be used to identify changes in bank accounts, categorize how suspicious they are, and identify fraud threats.

A variety of machine learning technologies are commonly referred to as artificial intelligence (AI) in fraud detection. An algorithm is a set of rules that are applied to a scenario in order to produce a decision in machine learning. Determine whether or not fraud is a threat in this situation.

The big benefit, though, is that these guidelines change as the technology gains knowledge. AI gradually improves its accuracy by "learning" from past situations.

Here are a few instances of machine learning technology:

- **Models for natural language processing (NLP):** Enhance and arrange data into more specific categories
- **Captcha and reCAPTCHA:** automated tests to distinguish between machines and people
- **Graph neural networks (GNNs):** a data processing system that illustrates the connections between various data points to enhance comprehension of the whole.

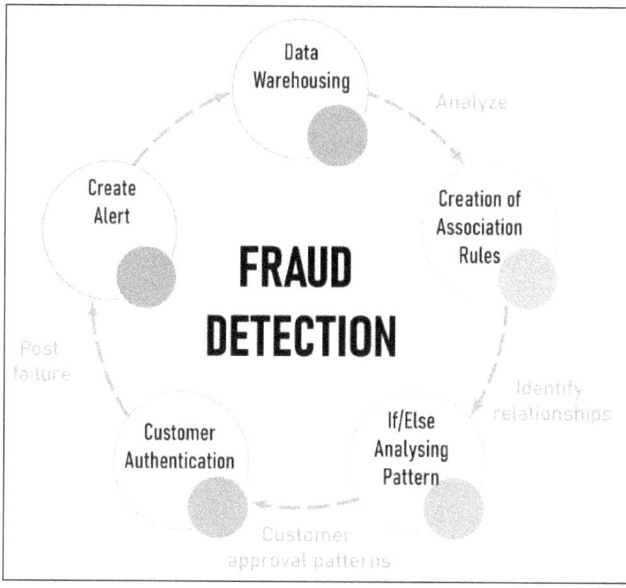

***Figure 4.2** Fraud detection.*

Source: (arkoselabs, 2024)

1. Data Warehousing: Collecting and Storing Transactional Data

Data warehousing is the first step of the fraud detection process where large volume of financial transaction data is collected and stored. Customer transactions, payment histories, behavioral patterns are the type of data in this data. This data is fed into AI powered systems which detect any irregularities and help organizations establish such a baseline of what is normal financial activity. Data warehousing gives proper structure to real time and historical transactions, and hence proper fraud detection is possible.

2. Creation of Association Rules: Identifying Transaction Relationships

After the data is collected AI and machine learning algorithms come up with association rules to find the patterns and relationships of other transactions. This will help in finding fraudulent behaviors by linking an unusual activity to a known fraud scheme. For example, if the customer's transaction is similar to fraud patterns associated with it, the system triggers further review. As we all know, Anomaly and Fraud detection at scale is the most important demand of Association rule mining.

3. If/Else Analyzing Patterns: Setting Fraud Detection Conditions

AI models in analyzing patterns in order to apply logic-based conditions to see whether a transaction is legit or fraudulent. This means creating rules from previous fraud cases such as huge amounts of transactions at one time, many transactions in a small-time frame, or a different spending pattern than what is normal for a particular customer. The financial institutions continuously refine these conditions to increase accuracy of fraud detection and minimize false positives.

4. Customer Authentication: Verifying User Identity

The system allows the review of transaction patterns, leading to customer authentication using the AI-based verification of identity, for example, MFA or biometric verification, to verify that the transaction has been triggered by an authenticated user. This step substantially eliminates fraud through blocking of unauthorized account access and account owners confirming the legitimacy of customer actions. The authentication is also boosted by behavioral analytics, which include things such as keystroke dynamics and tracking of locations.

5. Create Alert: Notifying Suspicious Activity

The system will trigger an alert if a transaction is deemed fraudulent and will alert customers, fraud prevention teams and financial institutions. SMS notifications, emails or system logs that trigger immediate investigation are considered as alerts. AI driven fraud detection systems categorize alerts by severity, so that high risk threats can get quick, action and other transactions can be kept to a minimum.

6. Post-Failure Analysis: Enhancing Fraud Prevention Strategies

Finally, organizations analyze past fraud incidents in the post failure phase of the fraud detection cycle to perfect the AI models as well as the detection rules. Business, by the way of analyzing fraud pattern and the smart attack vector can update fraud prevention policies and also take precaution to offer more security. This step is very important as this helps ensure that financial institutions always stay at the forefront of emerging fraud tactics, allowing them to stay up to date with fraud detection efficiency and accuracy.

Integration of these six stages will help financial institutions to build a robust AI based fraud detection system that would proactively find the threats, reduce financial losses and increase customer trust in secure digital transactions.

4.2.2 AI-based risk scoring for financial transactions

Evaluating and analyzing possible risks connected to a certain financial decision, transaction, or investment is known as risk assessment in the finance industry. It entails estimating the probability of commercial, political, or economic shifts that might have a detrimental impact on a company's bottom line. Businesses can manually assess this kind of risk by looking at information and data. However, businesses can use AI to generate more accurate outcomes. Businesses can use this approach to make well-informed decisions about financial transactions, including stock market trading, insurance policies, investments, loans, and savings. In the end, applying risk assessment methods helps businesses and individuals to lower losses and boost returns in any financial sector.

Although it began to gain traction in the 1990s, AI-based financial decision automation made little headway over the following 20 years. But since the early 2010s, artificial intelligence (AI) and computer technology have made it possible to create increasingly intricate economic models and identify anomalies in data sets. In order to manage risk, keep an eye on transactions for possible fraud, and update credit scoring models, banks and other financial services providers are turning to AI algorithms. This improves efficiency and boosts profits at a low cost (Ellen Cibula, 2023).

The sheer volume and complexity of financial transactions continue to be on the rise, and the risk getting assessed for each transaction must be done in an effective manner; otherwise, it will affect an organization's security and operational efficiency. Traditionally, a fraud detection system uses static rules or predefined risk factors to examine the likelihood of fraud, which may become outdated in the case of dynamic digital finance. To overcome this problem, the approach of AI-based risk scoring is more sophisticated and adaptive to evaluate the degree of fraud in financial transactions.

Machine learning based on AI is used to evaluate a broad category of variables related to each transaction in order to determine the risk. While traditional scoring methods may only take into consideration elementary factors, such as the amount of transaction or where the account holder lives, AI-based scoring systems encapsulate a much wider amount of data. Such datasets include historical transaction patterns, device fingerprints, behavioral biometrics, geographical location, time of day, transaction frequency, and others. When thinking of these many data dimensions simultaneously, an AI system can understand a more complex and accurate risk score for each and every transaction; the likelihood that it is fraudulent.

The ability to process real time data also helps risk scoring models to deliver immediate insights into risk of a particular transaction. This differs from a manual review process, which can slow down the response and increase the period when a financial institution is open to fraud. Risk score (generally on low to high gradation), which signifies the chance that the transaction is fraudulent, is assigned by the system. Immediate action, such as manual investigation and more steps of verification or auto-rejection, is flagged for higher-risk transactions.

The key advantage of risk scoring derived from AI is that it is scalable. Today, millions or even billions are now being processed in a day by financial institutions. The ability of AI systems to scale tightly to evaluate millions of transactions at once while simultaneously scoring risk against transaction models in real time means there is no need to sacrifice performance in such systems. This is to ensure that fraud detection effort continues to be robust and still effective as the transaction volumes grow. It relieves the human investigators of the burden by devoting their attention to those high-priority cases that should be most likely to be cases of fraud.

Additionally, by utilizing AI-based Risk Score, false positives — transactions that are flagged by the fraud filter as fraudulent but are not — at least reduce in comparison. Unfortunately, these fraud detection systems often fall short when it comes to this problem, leading to unnecessary account freezes and an annoyance to customers. However, AI-based systems can take note of each transaction history and customer activity to make the generated risk scores more accurate. However, understanding the particular context in which each user's activity occurs makes it easier for AI to recognize abnormal behaviour in a manner that will cause as little inconvenience to customers who are legitimate transactors as possible.

Besides improving fraud detection, the use of AI based risk scoring is a major factor in helping the banks satisfy Know Your Customer (KYC) and Anti Money Laundering (AML) regulations. With the help of this art of AI, the institutions are being compliant with these regulations by flagging suspicious activity so they can be further investigated and rested. Reaching completely into risk management and legal compliance by using the combination of AI risk scoring and other compliance products, such as transaction monitoring systems and regulatory reporting.

4.3 AI for Anti-Money Laundering (AML) Compliance

The fight against financial crimes like terrorism financing and drug trafficking has grown much more important for the topics of anti-money laundering (AML). Rapid advancement in the technology has made artificial intelligence (AI) in AML more important. There is potential for AI to transform AML efforts into an accuracy and efficiency sandbox for detecting and stopping suspicious activity. At the same time, however, the use of another tech in AML also throws up some challenges and ethical considerations. Read this blog where we discuss the changes which AI has brought about to the threat of AML, its advantages, challenges, case studies and ethical implications. To grasp AI's impact on AML, one should understand the details of money laundering. Typically, money laundering takes place in three stages, placement, layering, and integration.

In placement stage, these illegally obtained funds are injected into the financial system, usually through smurfing (breaking down big sums into small, less suspicious amounts) or purchase of high value assets. At the layering stage, these funds are then layered through several transactions to make their origins untraceable, sometimes through sending money across borders or converting them into different currencies. In the third and final stage of integration, the "cleaned" money is put back into the economy, seeming legitimate funds to be spent on real estate investments, luxury goods, or legitimate business.

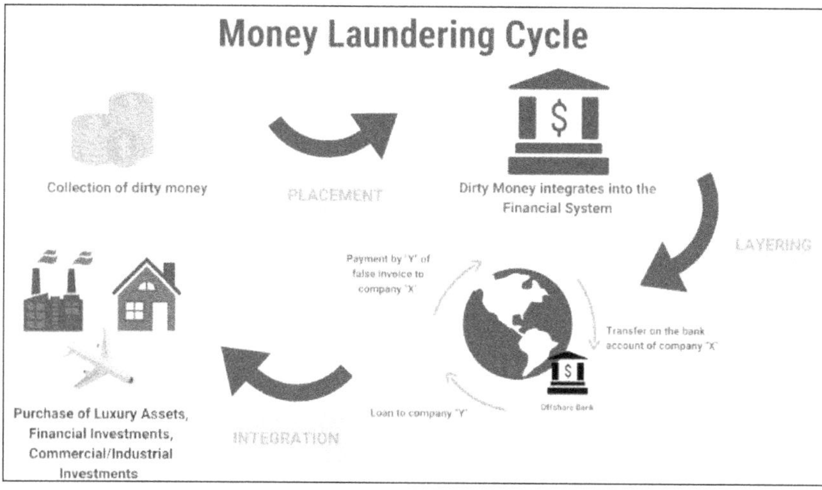

Figure 4.3 *Complexity of Money Laundering*

Source: (Linh Chu Dieu, 2024)

These operations are becoming increasingly complex and global in nature, and traditional AML processes involving lot of manual checks and predefined rules are becoming increasingly inadequate. These systems are slow, easy for a criminal to bypass, and overwhelmed by the amount of data that each criminal generates, which allows sophisticated criminals to circumvent the detection.

Role of AI in Anti-Money Laundering

Real time analysis of big data using AI algorithms is possible and it is capable of identifying underlying patterns and also pointing out unusual transactions or activities that require further investigation. Automating these tasks allows AI to increase the accuracy and speed of AML efforts and releasing those resources for the financial institution to focus on what's important. Apart from being more efficient, AI can also be more effective in AML efforts. AI can keep discovering new patterns and trends that in turn helps in identifying the risks which are not being studied by the current processes and furthermore enhance the accuracy of the AML programs.

In short, AI will increase sophistication and automation in the AML process to detect and prevent financial crimes, as well ultimately taking a burden off of financial institutions. Nevertheless, there are also issues both in terms of challenges and ethical considerations when applying AI in AML.

Advantages of AI in Anti-Money Laundering

Advantages of using AI in AML efforts are several.

1. **Accuracy:** AI can assess and analyze humongous amounts of data effortlessly and in real time to find out the patterns which may not be detectable by humans. This lets you, as financial institutions, find suspicious transactions and behaviors and lower the false positives.
2. **Reduced resources:** With some of the manual tasks related to AML became automated by AI, the resources are able to be redirected to other important AML tasks. It also speeds up availability time to address a potential threat given that financial institution and reduces the time already spent on compliance reviews.
3. **Adaptability:** AI can adapt to new patterns and trends and continuously learn; thus, financial institutions can maintain its position by staying ahead of the curve in terms of new threats and risks.
4. **Cost efficient:** Financial institutions are able to cut costs incurred from compliance review and investigations due to automation of many AML processes.
5. **Customer experience improvement:** AI helps to motorize and simplify the customer due diligence procedures, thus, making it easier and faster for customers to create accounts and do transactions.

Challenges in implementing AI in Anti-Money Laundering

Many advantages of using AI for AML but there are many challenges that financial institutions experience when implementing AI in their AML programmes.

1. **Data quality:** The quality of the data that the AI algorithms work with is paramount to the extent that they **work**. Nevertheless, data quality issues may be an impediment for financial institutions as data may be incomplete or inaccurate. This can result in false positive or false negative cases or reduce the effectiveness of AML programs.
2. **Participating ability:** Financial institutions need to be able to participate in complex rules that AI algorithms can decide, but that is sometimes hard to explain to regulators or auditors how the decision was made.
3. **AML regulations:** Financial institutions need to comply with the very complex and increasingly changing AML regulations. While AI can be implemented in AML programs, there may be a lot of **changes** needed to the current processes and systems which can be challenging to keep the compliance.
4. **The ability of AI to automate AML processes:** Although AI can automate many AML processes, one cannot exclude AI from decisions made after AI insights. It is not easy for financial institutions to **find** the qualified people who can do the work using AI to enhance AML efforts.
5. **Bias:** AIs can be biased for two reasons: 1. If the data used to train an AI is biased and 2. If the AI is not built in a manner that addresses biases. Such discrimination can result in unfair and potentially biased predictions.

Importance in various industries

In many industries, anomaly detection is a valid concept and therefore, it is a versatile and important concept:

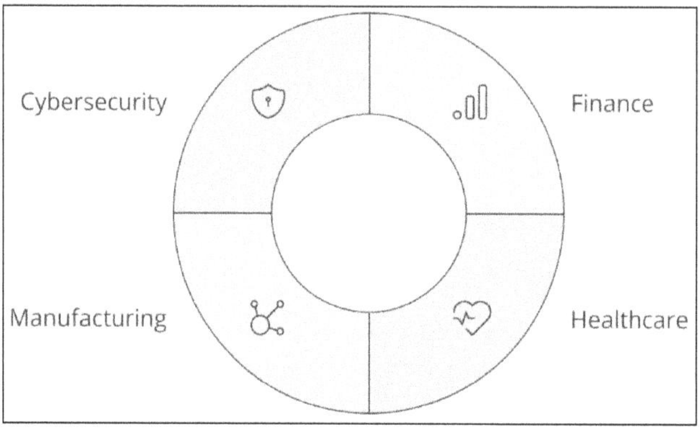

Figure 4.4 *Anomaly Detection Importance.*

Source: (Victoria Shutenko, 2024)

- **Finance:** AD (anomaly detection), the system in the finance sector, help identify fraud transactions and unusual trading patterns to prevent financial losses and upholding the regulation of the sector.
- **Healthcare:** AD techniques can provide benefits to healthcare by identifying unusual patient records that may indicate a health issue or an incorrect entry in the data, increasing patient care and operational efficiency.
- **Manufacturing:** Real time monitoring of equipment for anomalies can predict and prevent potential failures in manufacturing, saving great costs and preventing downtime.
- **Cybersecurity:** AD provides the ability to detect unusual network activity and prevent security breaches and maintain system integrity and protect vulnerable data.

Case Studies: AI in Anti-Money Laundering

Various successful case studies have been made in using AI in AML efforts. Here are a few examples:

- **Danske Bank: Enhancing AML Accuracy and Reducing False Positives**

In 2018, Danske Bank launched an AI-based anti-money laundering (AML) system to improve its fraud detection capability. The advanced system which used machine learning based algorithms to analyze in real time huge amount of customer transaction data to find patterns and anomalies and to detect suspicious activities. This enabled the bank to employ AI to separate the valid from the fraudulent transactions better. AML compliance has a high rate of false positives, i.e., the rate of false positive transactions. Danske Bank has successfully decreased the number of false positives by 60 per cent based on the use of the new AI system, which has greatly improved the efficiency of the operation. By allowing compliance teams to focus on only the most high-risk transactions, it increased effectiveness of an AML program as a whole.

- **HSBC: Automating AML Processes for Cost and Time Savings**

In 2019, HSBC took a great stride during the launch of the automated AML system driven by AI to automate its compliance processes. Machine learning algorithms that the bank has used to analyze customer transaction data are used to detect suspicious behaviours of money laundering and financial crime as soon as possible. In most cases, AML methods have been traditionally manual reviews, which is challenging and time consuming. AI was integrated by HSBC to streamline AML reviews and reach faster and more accurate risk assessment. The manifestation of these financial benefits of automation in compliance operation was the $400,000 per year of efficiency improvements made by adaptive design of the system. Not only it had contributed to making the bank more secure, but it also improved accuracy and thus made the bank compliant with global regulations with less operational burden.

- **Standard Chartered: Reducing Compliance Review Time with AI**

In 2020, Standard Chartered brought in an AI based AML solution that eliminates the need of human action to automate fraud detection and compliance reviews. Real time transaction data was processed by the system and in real-time high-risk transactions were identified with greater accuracy. Using machines to do the work of their employees, the bank did away with 40% of the time spent on compliance reviews and saw dramatic improvements to the efficiency of its workflow. Using the customer behaviors and transaction networks, the AI system could analyze suspicious activities more effectively and find links between them than traditional rule-based methods. Standard Chartered reduced the burden on human compliance teams and focused on the riskiest cases so that its AML efforts remain regulatory compliant and accurate. The implementation demonstrated how AI can be used to optimize operational processes of financial institutions and thus the speed and reliability of AML programs.

- **JPMorgan Chase: Minimizing False Positives and Enhancing AML Precision**

In 2021, JPMorgan Chase made a major foray into deploying an AI-powered AML system that relied on machine learning to tackle fraud cases. The system was created to examine huge volumes of customer transaction data and to detect activities that would be unusual and indicate potential money laundering. A breakthrough of this implementation was a reduction by 95% in false positives one of the greatest improvements of the traditional AML systems. Running the new model would significantly reduce the number of mistakenly flagged transactions, enabling the bank's compliance teams to concentrate on actual risks and thus enhance fraud detection and the overall AML program accuracy. It also helped JPMorgan Chase comply with the latest financial regulations while streamlining operations through the use of the AI driven system.

4.3.1 AI-powered anomaly detection in financial networks

The process of anomaly detection in financial networks is a very important task that allows the banks, payment processors as well and regulatory bodies to detect potentially fraudulent transactions in real time. Artificial intelligence (AI) and Machine Learning (ML) make it possible for financial institutions to process huge amounts of transactional data, flag abnormal behaviour as well and act early to stop financial crimes. Unlike the classical method of fraud detection that relies only on fixed rules, AI-based anomaly detection continues to learn from historical data and learn according to new fraud techniques, which makes it able to detect emerging fraud and adapt new fraud techniques.

Financial networks are made up of a set of entities – individual customers, businesses, and so on, to payment gateways, international institutions and so on. Anomalies appear in different ways in these networks, for example a sudden spike in the number or value of transactions, strange cross-border payments, several rapid transfers of funds between different accounts or different behavior for a given customer. These patterns serve as seed points for the AI-driven anomaly detection systems that are analyzing them in real time, trying to identify inconsistencies that may be referencing fraud, money laundering, malicious insider attacks, cyber-attacks, etc.

The integration of AI in anomaly detection means increasing financial security by reducing false positives and giving a better understanding of suspicious activity. Traditional systems have played this role of generating too many alerts, and therefore, they are too inefficient and end up being costly for the compliance teams. However, AI-based models further improve their ability to differentiate between real deviation and malicious anomaly; no high-risk transactions should be missed out, and no true legitimate customers will be disrupted unnecessarily.

Data anomaly detection is the process of identifying some points (or transactions, or patterns) that deviate from the norm. Anomalies of this type may suggest possible fraud, error, inefficiency and other irregularities in financial transactions which should be immediately drawn to the attention of the authorities.

Importance of Anomaly Detection in Financial Transactions

Financial loss and noncompliance with regulations can result from even the smallest irregularity in transaction data. Strong anomaly management is even more important as the amount of digital transactions rises. The following justifies the significance of anomaly detection in financial transactions:

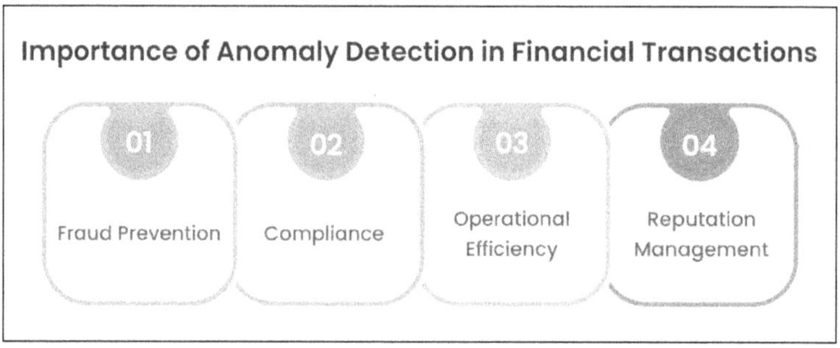

Figure 4.5 *Importance of Anomaly Detection in Financial Transactions.*

Source: (HighRadius, 2024)

1. **Fraud Prevention:** Preventing financial losses due to fraud requires early identification of unusual trends in financial transactions. Businesses can prevent serious financial harm and preserve the integrity of their financial systems by spotting and resolving these irregularities before they become more serious. In actuality, the Anomaly Management software from High Radius uses AI-based algorithms to retrieve 12 to 18 months' worth of historical data from your ERP and examine it for trends. If some transactions are not in keeping

with patterns, they are flagged as anomalies in worklists with suggestive actions for resolving the anomalies. Financial reports are considerably more accurate and closer much faster thanks to the speed of AI powered detection.

2. **Compliance:** Businesses are obligated to keep up with relevant records of their finances in order to comply with the stipulations set by regulatory bodies. In this, anomaly detection is very important as it calls out any differences in transaction data so business can correct them early and for the most part avoid any penalties or worse, legal consequences.
3. **Operational Efficiency:** Businesses can quickly detect errors in financial transactions such that these errors do not compound and disrupt financial operations. This proactive approach is not only for increasing operational efficiency, but also for making the financial processes run smoother and without disruptions.
4. **Reputation Management:** In the financial sector, this can be paramount; therefore, all the way to having a good reputation for themselves is what's good for them in the long term. With the ability to detect and fix anomalies effectively, businesses can maintain data integrity and show their dedication to security and reliability in order to ensure the reputation and trust of customer.

Techniques for Anomaly Detection in Transaction Data

The methods used for anomaly identification vary from sophisticated machine learning and deep learning algorithms to traditional statistical methods. These methods aid in accurately identifying outliers that are the consequence of fraud, mistakes, or a variety of other anomalies. Financial institutions can protect the integrity of their transaction data and take preventative measures against attacks by being aware of the best detection techniques and knowing how to use them. Let's examine a few anomaly detection strategies:

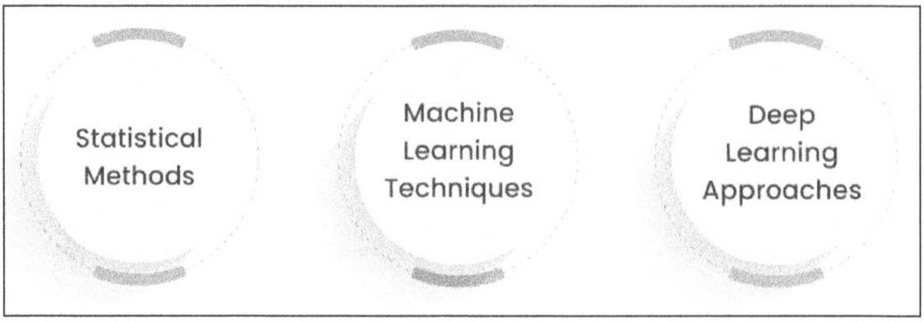

Figure 4.6 *Techniques for Anomaly Detection in Transaction Data.*

Source: (HighRadius, 2024)

1. Methods of Statistics:

- **Z – score:** The Z-score is a useful tool for identifying outliers in normally distributed data since it quantifies the number of standard deviations an observation deviates from the mean.
- **Moving Average:** This method smoothes out short-term swings and identifies long-term patterns or trends; it highlights anomalies more clearly than they would be if they were shown in their unprocessed state.

2. Methods of Machine Learning

- **Clustering:** Similar data points are grouped together in unsupervised learning approaches like ask-means, while outliers create their own clusters.
- **Classification:** Supervised learning models may also be used to classify transactions into normal and abnormal categories.
- **Neural Networks:** Neural networks would be the ideal choice for identifying such complex abnormalities in vast, complex datasets.

3. Deep Learning Methodologies

- **Autoencoders:** Neural networks called autoencoders are made to reconstruct input data. They are adept at drawing attention to departures from the norm in order to spot anomalies.

- **LSTM Networks:** This type of model makes it simple to look for anomalies in sequential data, including time-series financial transactions.

4.3.2 Automating AML reporting and suspicious activity monitoring

The goal of anti-money laundering (AML) policies, rules, and regulations is to stop criminals from committing financial crimes and engaging in other unlawful activities. To stop financial crimes and criminal activity, both local and international authorities are set up all over the world. These regulators create policies. Despite the fact that compliance can be difficult, businesses must abide by these rules. Consequently, financial institutions purchase software and have compliance divisions.

Because they are required to notify authorities of any suspicious activity, financial institutions are the most common users of AML laws. Financial institutions must do more than just report questionable activity, though. However, because they give credit to customers who open accounts with the business, they are also more vulnerable to money laundering.

Significance of AML Compliance

According to a conservative estimate, the amount of money laundered internationally in a single year is between 2% and 5% of the world's gross domestic product, or $800 billion to $2 trillion. Illegal arms sales, smuggling, embezzlement, insider trading, bribery, and computer fraud schemes are often linked to money laundering. It is also common in organised crime, including prostitution rings, the trafficking of firearms or drugs, and human trafficking.

Financial institutions use counter-funding of terrorism (CFT) to stop the funding of terrorists, and it is closely related to anti-money laundering. AML laws cover both money laundering (the source of funds) and terrorism funding (the destination of payments).

Financial institutions employ AML strategies for the following reasons, in addition to their moral obligation to prevent money laundering and terrorism financing:

- They must keep an eye on customers and transactions in order to comply with regulations, and they must report any suspicious activity.
- Preservation of the value to shareholders and the reputation of their brand.
- Cost savings from fines, staff, IT expenses, and money set aside for risk exposure.
- Preventing consent orders and possible civil and criminal penalties for negligence or noncompliance.

Types of AML Transaction Monitoring

To identify and stop fraud, money laundering, and other financial crimes, businesses and financial institutions can employ a variety of transaction monitoring techniques. Depending on the needs of the organization, these techniques differ in terms of complexity, automation, and efficacy. Typical forms of transaction monitoring include the following:

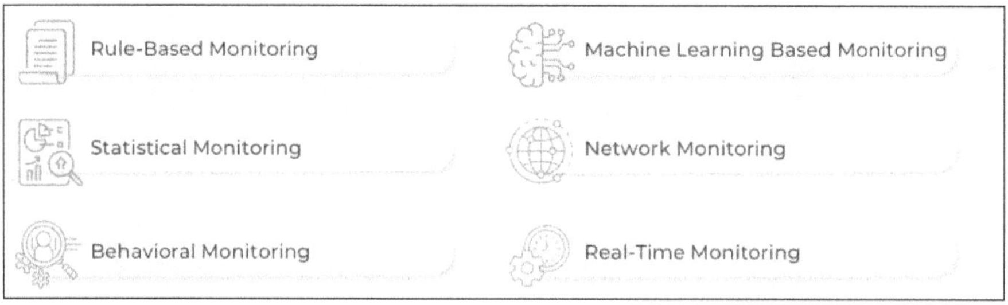

Figure 4.7 *Types of transaction Monitoring.*

Source: (Rohit Rao, 2024)

1. **Rule-based monitoring:** It involves setting up pre-defined rules and thresholds that trigger alerts when transactions meet specific criteria. These rules can be based on transaction

amount, frequency, location, or other factors that indicate potential risks.

An example of this is the rule-based monitoring done by a bank to monitor for suspicious activity on the account of a given customer. The bank's security team is alerted to investigate if the amount of a transaction is above a certain dollar amount, if the transaction takes place in a location far from the customer's typical spending patterns, or at an unexpected time of day. The proactive approach enables the bank to rapidly recognize and respond to possible fraud or theft before a costly stake in the customer or institution is suffered.

2. **Statistical monitoring:** Mathematical models and algorithms are used to analyze transaction data and look for patterns or abnormalities which could indicate that the behavior is fraudulent in statistical monitoring. It is more advanced than rule-based monitoring as it can internally change with the trends and patterns.

   For an example, a statistical monitoring can be used by credit card company to identify strange spending on a customer's credit card account. They can use to analyze their transaction history and compare with the strong set norm distribution or behavior to declare their transaction aren't appropriated but are malicious. But with this advanced approach the company can respond quickly to new threats and new fraud tactics in real time.

3. **Behavioral monitoring:** Behavioral monitoring focuses on the analysis of the behavior of individual customers or entities in order to find deviations from the usual activity pattern. Creating a baseline of normal behavior per customer for each business, allows them to see any unusual or suspicious transactions which may indicate fraud or money laundering.

As an example, a cybersecurity firm may run a behavioral monitoring system to find out possible attacker inside a company. The system analyzes employees' normal ways of getting to sensitive data or systems and flags any odd behavior, like being on at an odd hour or trying to download lots of data in a short span of time. This allows the firm to be proactive in screening suspicious activities and prevent data breaches or unauthorized access well before they reach the highest level of a security incident.

4. **Machine learning-based monitoring:** Transaction data can be analyzed by algorithms who learn what has taken place in transactions in the past in order to find patterns and anomalies. As the machine learning models can learn and adapt over time, they are very powerful detectors of complex fraud schemes that other monitoring schemes may not spot. However, this type of monitoring requires a lot of data and computation resources, but it will deliver more accurate and timely results.

   For an example, machine learning based monitoring can be used by an e-commerce platform to identify fraudulent activities by its customers. The system analyzes huge volumes of transaction data and learning from previous trends, is able to discern tiny indications of fraud singularities that you may miss if you are a human and even from an ordinary rule-based systems. The platform's advanced algorithms keep improving their understanding of fraud indicators so that the platform can proactively prevent fraudulent transactions and preserve the business from the financial losses and its customers.

5. **Network monitoring:** It means analyzing the relationships between different entities in a financial network to find out possible illicit activities. When in place, network monitoring

can assist in tracking suspicious transactions for the purposes of spotting a money laundering or fraud scheme›s overall flow of funds between accounts and institutions such as banks.

For instance, a global financial network may be monitored by a network for tracking down money laundering activities in a government agency. The system analyzes the complex interrelationships between different entities and also how funds are moved from one account to another and in many institutions, for instance it can reveal suspicious transactions that could be associated with crimes like drug trafficking or terrorist financing. By trailblazing the financial trail, this method enables authorities to snoop out elaborate financial crimes and bring down criminal networks as a whole.

6. **Real-time monitoring:** Real time monitoring means examining transactions while happening in real time to locate and react to any things that are dubious. Finding and preventing fraud before the losses are substantial is critical, and this type of monitoring is very important.

   For an example, a real time monitoring may be performed by a payment processing company, as it can quickly detect fraudulent transactions in order to protect its customers from financial losses. From there that monitoring platform can instantly flag any suspicious activity from unauthorized purchases to unusual spending patterns when that data is constantly being scrutinized from within the flow system. Such a rapid detection and response capability allows the company to stop fraudulent transactions in real time, prevent the occurrence of any risks in the first place and protect both the customer's and the company's financial interests.

Benefits of AML Automation

For companies and financial institutions, AML automation improves the efficacy and efficiency of AML compliance operations. More significantly, automating AML procedures lowers the risk of reputational damage while also reducing the resources needed for compliance operations.

Implementing AML automation in additional areas can assist businesses in a number of ways, including:

Figure 4.8 *Benefits of using an automated AML software.*

Source: (Gabija Stankevičiūtė, 2023)

1. Easier Risk Assessment

Advanced machine learning algorithms further augment automated AML software, allowing for the use of high number of advanced algorithms on massive amount of transactional data in real time for more sophisticated risk assessment. Automation allows for a more precise customer behavior, financial transaction and suspicious activity evaluation while traditional manual risk assessment methods can be time consuming and prone to errors. Continuously modelling risk, the system classifies individuals and organizations as being high risk, allowing the financial institutions to take action against the potential fraud and money laundering activities.

2. Optimized Compliance Workflows

Financial institutions have different stages of compliance workflows such as CDD, transaction monitoring, and regulatory reporting. Using automated AML software, the task of these workflows is streamlined integrating rule-based decision-making process with artificial intelligence driven insights. As a result, it leads to a compliant management system that minimizes human involvement and yet delivers regulatory standards in an efficient manner. By automating workflows, redundancies are removed from the systems and accuracy is increased, without collapsing into compliance-related bottlenecks.

3. Robust Screening and Transaction Monitoring

AML software performs real time screening and monitoring of financial transactions against global watch lists, sanction lists and politically exposed persons (PEPs) is one of its critical functionalities. AI can help automated AML solutions track suspiciously active patterns of activity which, in turn, can assist them in getting in front of money laundering efforts before they progress further. Through this continuous monitoring institutions get to know if there are any emerging financial threats so that they can always be a step ahead of the threats and also keep up with the local and international regulatory bodies.

4. Simple Data Collection and Analysis

Financial institutions, however, are in need to gather and analyse large amounts of customer and transaction data in order to comply with AML. As such, automated AML software helps simplify this process with the help of centralized data collection and AI based analytics. This speeds up compliance team's ability to quickly evaluate large datasets, find anomaly and create insights to improve fraud prevention. Structured and automated data management of information helps the organization in complying with the audit requirements and reporting obligations.

5. Efficient Data Reporting and Audit Trail

Under the regulatory authorities, financial institutions must keep a clear and transparent audit trail of all the compliance related activities. Automated AML software allows for a systematic reporting mechanism where everything, from a transaction, risk assessment and compliance action is logged systematically. It produces detailed reports for the audits which makes it simpler for institutions to show that they are complying with AML regulations. Automation also contributes to reduce errors and inconsistency in compliance documentation.

Reduced Costs and Ensured AML Compliance

The cost reduction is one of the most important benefits of implementing automated AML solutions. High operational costs of manual AML compliance processes are due to the need of large teams of compliance officers. Through automation, financial institutions are able to cut labor costs to a large extent, while increasing the efficiency and effectiveness of AML compliance efforts. Organizations also know that automated systems will ensure that regulatory requirements are met promptly, so that hefty penalties aren't incurred for noncompliance.

AML software using Artificial Intelligence (AI) churning enables raising the financial security at large as well as in compliance with the regulation. That allows banks to have more accuracy in detecting suspicious activity while lowering errors and compliance bottlenecks. These solutions simplify data collection, improve reporting efficiency, and store auditing trail which makes institutions meet transactional requirements seamlessly. Also, automation greatly reduces operational costs by reducing the reliance on large compliance teams as well as the delay in following financial regulations. Automated AML software is overall faster, more precise, and more cost effective than preventing financial crimes and at the same time complying with the regulations.

4.4 Behavioral Analytics for Financial Security

With the rise of digital and sophisticated financial crimes, protecting legitimate financial transactions is no longer sufficient through traditional security measures such as securing passwords and fraud detection based on rules. Fearlessly, criminals continuously develop new tactics to breach security systems, which calls for financial institutions to adopt more advanced methods to figure out what is fraudulent. In this regard, behavioral analytics helps in improving financial security.

Financial security through behavioral analytics relies on AI, ML, and big data analytics to examine the behavior patterns of users and seek for anomalies that may mean risks of fraud or bad action. Instead of learning from static rules or limits on transactions that limit the people based on my set criteria, behavioral analysis takes the approach of learning about how customers interact with financial platforms, not only at the moment but over time, thus creating a unique behavioral fingerprint for every customer.

By using this methodology, banking institutions can detect aberrations in the activity of the user, like strange login locations, fast fund transfers, odd transaction amounts, or abrupt alterations in spending practices. Behavioral analytics utilizes the keystrokes, mouse movements, device usage, or transactions of users to figure out if legitimate users are conducting transactions or, in the instance of fraud, an unauthorized entity. Say, a customer, who traditionally makes small, domestic purchases goes ahead and initiates large international wire transfers, the system will be alert and respond as the potential risk.

Challenges in Behavioral Analytics Detection

It can be challenging to develop a machine learning system that can identify unusual activity and raise red flags. The following is a list of difficulties in detecting behavioural fraud:

Table 4.1: Challenges in Behavioral Analytics Detection.

Challenges	Description
Variability and Quality of Data	Inconsistencies in data sources, formats, and quality can compromise fraud detection systems' precision and efficacy.
Behavioural Patterns' Complexity	Accurately comprehending and interpreting behaviours in the constantly changing fraud landscape can be difficult.
Negative and False Positive Results	One of the biggest challenges is balancing detection accuracy to reduce false warnings without missing real fraud cases.
Integration Difficulties	Technical limitations, data silos, and compatibility problems could prevent smooth integration.
Challenges in Identifying Novel Fraud Techniques	It is quite difficult to identify new fraud trends and modify analytics models appropriately.
Data Privacy Issues	protecting user privacy while successfully identifying fraudulent activity.

Source: (Self-generated)

Benefits of Behavioral Analytics in Fraud Detection

Conventional security methods frequently fail to identify these intricate threats. Here, behavioural analytics shows up as a potent instrument with several advantages for preventing and detecting fraud. The following are the main advantages of behavioural analytics for fraud detection:

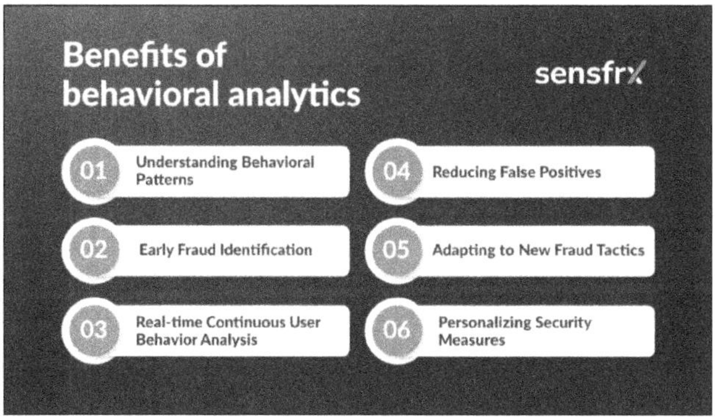

Figure 4.9 *Benefits of behavioral analytics.*

Source: (sensfrx, 2023)

- **Understanding Behavioral Patterns:** Behavioral analytics allows understanding of pattern of behavior so they can be established as baselines, and deviations continuously monitored to detect anomalies. For fraud detection, this approach is essential as one expert can miss a subtle pattern. For example, the fintech company at can analyze the login times, amounts of transactions and device fingerprints to detect the suspicious activity.
- **Early Fraud Identification:** Real time monitoring of behavior analytics allows organizations to detect some suspicious structures in time and take immediate measures and prevent fraud impact. An example for this is an e-commerce retailer able to monitor browsing patterns, cart abandonment and purchase history. An unusual deviation such as rapid multiple purchases or big transaction can be flagged as fraudulent.
- **Real-time Continuous User Behavior Analysis:** Behavioral analytics finds suspicious deviations immediately so that businesses can act in time and stop any financial loss. Let's say an e-commerce retailer monitoring people's browsing history, cart abandonment and purchases made: this could identify

a sudden increase in purchases, or a very big basket sizes as possible fraud attempt.

- **Reducing False Positives:** In a manner of speaking, traditional fraud detection systems generate lots of false positives, wasting resources. However, behavioral analytics, a part of machine learning, minimizes these errors. Imagine that training a system to identify a pattern of regular travel can then be used to recognize one type of fraudulent activity from another, e.g. legitimate from legitimate, or legitimate from fraudulent.
- **Adapting to New Fraud Tactics:** The game of frauds is ever evolving, but behavioural analytics leverages machine, learning as a way adapt and thrive. As a result, a cybersecurity firm can provide its models with training across emerging phishing techniques or malware signatures so they can better detect fraud.
- **Personalizing Security Measures:** Behavioral analytics can provide detailed user profiling that is useful to personalize security. In fact, a risk score may be based on an online banking platform user based on login location, transaction type, or device fingerprints. Extra authentication may be applied for high risk users like IP whitelisting or biometrics.

4.4.1 AI-driven customer behavior analysis for fraud prevention

With the rise in the number of digital transactions and the intricacy of digital payment transactions, financial institutions are left with the greater task in order to find ways to detect fraudulent activities and remain effective in delivering the experiences to the end customers. Most of the traditional fraud detection methods are based on predefined rules and static thresholds that easily can be bypassed by high sophisticated cybercriminals. On the flip side, AI-enabled customer behavior analysis is a dynamic and dynamic way to preempt fraud using machine learning, big data analytics, and behavioral biometrics to find out anomalies in real time.

AI-driven behavior analysis does continuous monitoring and learning from a customer's transaction history, device usage, spending habits, login pattern and interaction style. AI systems can also recognize deviations from normal behaviors and flag potentially fraud transactions as each customer is given a unique behavioral profile. Let's say a routine small, close to home customer initiates an enormous remote wire exchange, the AI framework can analyze as a deviation and request and further security runs before affirming the transaction.

The main advantage of AI-driven behavior analysis is that it can detect fraud in real time while minimizing false positives. Unlike 'rule of thumb' type systems, whereby warnings are usually called out for typical lawful transactions, AI models tend to have a wider view of behavioural aspects, allowing some degree of accuracy when determining what normal customer behaviour is from fraudulent conduct. Moreover, AI is continuously developed and improved through new data available to learn from and ways of adapting to evolving fraud techniques, hence making it an extremely good tool for financial security.

Also, there's no way for AI-powered fraud prevention systems to detect the higher sophistication of the fraud, such as account takeovers, synthetic identity fraud and automated bot attacks. Tricking AI (By analyzing keystroke dynamics, mouse movement patterns etc) even when the bot has stolen the credentials allows AI to check whether the credentials have been stolen or used by a human user or a bot, which will prevent unauthorized access. This helps security without being based only on traditional authentication means such as passwords or two-factor authentications, which may be compromised.

Limitations of AI-Driven Customer Behavior Analysis

1. **Privacy and Data Security Concerns:** Continuous monitoring of customers' activity dramatically increases privacy issues with a potential regulatory fallout, making AI fraud prevention function very complicated. Data protection laws (GDPR or CCPA) require that financial assets are compliant and transparent about how the customer data is used.
2. **Bias in AI Models:** If such historical data contains biased information, then the trained AI models could propagate the biases in their output, and thus, it will be discriminatory. For instance, an AI may encode such things in its code, resulting in the unfair targeting or the refusal of service to good customers based on transactions from certain demographic groups. Such biases have to be addressed using continuous monitoring and ethical AI frameworks.
3. **High Implementation Costs:** However, to build and embed fraud detection systems fueled by AI, the investment in infrastructure, personnel, and ongoing care is laborious. It may be difficult for small financial institutions to deploy and maintain AI-based security solutions as large banks can with large resources available at their disposal.
4. **Evasion by Sophisticated Fraudsters:** However, as AI can detect signs of fraud, cyber criminals always find ways to circumvent any form of detection. Low-volume, slow-burn attacks and even AI fraudsters may be used by fraudsters to simulate customers behaving normally. As a result, continuously improving and retraining of the AI models to keep up with its latest threats are required.
5. **False Positives and Customer Experience Impact:** AI indeed lowers the false positives compared to normal techniques but has errors that can flag legitimate transactions as fraudulent. It can be a frustrating experience for customers, and security can often be balanced with customers' convenience.

Understanding the Importance of Financial Behavior

In the modern era of personal banking and coordinated financial services, there is a growing necessity to comprehend how the customers behave in order to thwart frauds, manage risks and provide tailored banking services. Financial institution are interested in understanding customer behaviour patterns to detect anomalies, stop fraud, increasing security to better understand how can improve customer experiences. However, rising digital banking, mobile payments, and AI powered financial services can help in analyzing financial transactions to understand spending habits, transaction trends as well as the danger of such transactions being a result of fraudulent activities.

Various factors affect the behavior of customers in financial transactions, such as income level, lifestyle, geographic location, and purchasing preferences. On the other hand, the targets are set by the customers to achieve their financial goals. From these behaviors, banks and financial service providers learn how to develop personalised financial products, provide suitable recommendations, and make better risk assessment models. This process is further increased by AI-driven analytics that automated data collection, detect patterns, and predict financial behavior in the future with very high accuracy.

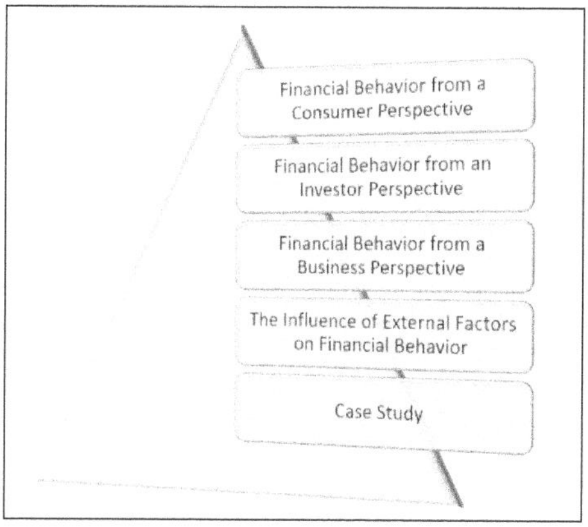

Figure 4.10 *Understanding the Importance of Financial Behavior.*

Source: (fastercapital, 2024)

1. **Financial Behavior from a Consumer Perspective:** Consumers' actions and choices have a huge impact on businesses when it comes to financial behavior. Knowing will help businesses cater to customer's needs and preferences. For instance, companies can analyse consumer spending patterns to find trends and take appropriate precautionary steps on how to reach the consumers.
2. **Financial Behavior from an Investor Perspective:** Financial behaviour is shaped by investors. Their decision to invest in a business can influence and lead a business to success or growth. This is because businesses can understand investor behavior and attract potential investors as well as secure funding for expansion or new ventures. For example, if there is a strong financial track record and a clear growth strategy, investors may find such an opportunity to be profitable.
3. **Financial Behavior from a Business Perspective:** Businesses themselves exhibit financial behavior through their financial management practices. Therefore, financial behavior should be effective including budgeting, cash flow management and investment decisions. By practicing sound financial, businesses are able to make the best use of their resources, avoid the burden of risks, and excellent profitability. For example, businesses can use cost cutting measures to protect themselves during period of lean.
4. **The influence of External factors on Financial Behavior:** Financial behavior is impacted highly by external factors as they include economic conditions, government policies, and market trends. It is important for businesses to keep informed about such factors and adapt their strategies to them. Things like change pricing, looking at new markets or add diversification in product offer might be required by businesses at a recession to stay competitive.

5. **Case Study:** Analysis of customer spending behavior helps the business identify a developing demand for environmentally friendly products. Feeling inspired to introduce a new line of sustainable products, they increase sales and gain customer trust. The example of this is that understanding and responding to the financial behavior can help business grow and succeed.

4.4.2 AI-powered biometric and multi-factor authentication

Each record in a blockchain is authenticated, and the events and balances created in the ledger can't be manipulated as it is shared by all participants in the network. Traditional passwords and PINs were often prone to security breaches, phishing and credential theft, as it was possible. To minimize these obstacles, there has been the rise of AI-powered authentication systems using machine learning, biometrics and behavioral analytics to provide secure authentication while keeping convenience in practice.

The AI-driven authentication systems are developed to analyses real-time data, to look for anomalies, and to operate in changing security threats. It is quite different from a static password-based system where users are required to remember complex credentials and while an AI authentication system does not require users to remember any credentials as they learn from user behaviors and device interactions along with context to provide dynamic and personalized security based on that. These are systems which anti identity fraud, anti-account take over and anti-unauthorized access in banking, healthcare e commerce and other sensitive domains.

The Evolution of Biometric and Multi-Factor Authentication (MFA)

Significant advancement of authentication methods occurs because cybersecurity threats continue to evolve, thereby ensuring secure access to sensitive systems and financial transactions. Users must now replace traditional password authentication because the increasing incidents of credential theft and phishing attacks, together with identity fraud, render passwords inadequate. As a result, biometric authentication and multi factor authentication (MFA) have arrived as powerful security alternatives, which provide better security while continuing the convenience of the user.

Yet, biometric authentication and MFA have also been stimulated by artificial intelligence (AI), machine learning, and sensor technology advancements to drive the development. Users can establish secure identity verification through authentication methods that combine built-in (biometric), memory-based, along possession-based features.

Biometric authentication leverages **unique biological traits** to verify identity. Unlike passwords and PINs, which can be stolen or forgotten, biometric features such as **fingerprints, facial recognition, iris scans, and voice recognition** provide a more reliable and user-friendly means of authentication.

Benefits of Implementing Biometric Authentication

Implementing Biometric Authentication in mobile applications offers several important benefits:

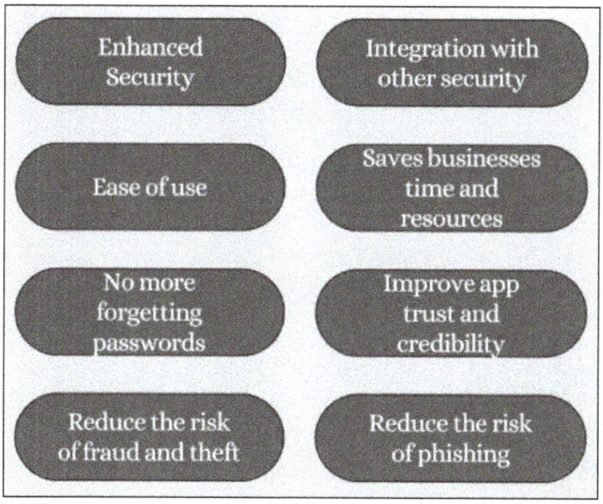

Figure 4.11 *Benefits of Implementing Biometric Authentication.*

Source: (Splendapp, 2024)

- **Enhanced Security:** The level of security offered by biometric authentication is far higher than that of conventional passwords. Because biometric data is unique to each person, it is very difficult, if not impossible, to fake or steal.
- **Ease of use:** Fingerprint or facial recognition is much faster and easier than a password. This makes it easier for the user and speeds up access.
- **No more forgetting passwords:** Biometric authentication relies on physiological trait, so users do not have to remember multiple passwords, which decreases the chances of forgotten passwords or resetting them numerous times.
- **Reduce the risk of fraud and theft:** The biometric authentication makes it more difficult for fraudsters since they need the user's real biometric data in order to access their personal accounts.
- **Integration with other security technologies:** Biometric authentication can be integrated with other security technologies, such as encryption and multi-factor security, to provide comprehensive protection.

- **Saves businesses time and resources:** Reduces the need for technical support for password issues and improves the overall efficiency of security systems.
- **Improve app trust and credibility:** Mobile apps that use biometric authentication demonstrate a commitment to security, which enhances users' trust in them.
- **Reduce the risk of phishing:** Phishing techniques aiming to fool passwords are difficult to pull off against biometric authentication.

Multi-Factor Authentication (MFA)

Multi-factor authentication is an authentication method that requires the user to authenticate using two or more factors or verifying factors to prove their identity before granting access to a resource such as an application, online account or VPN. MFA is a fundamental part of a solid identity and access management (IAM) policy. MFA asks for a username and password, but instead of just that, it asks for one or more additional verification factors, in order to decrease the chance of a successful cyber-attack.

Key Components of Multi-Factor Authentication

Different factors are the key components of an MFA system, along with the tokens used to enable each factor. There are specific software or hardware requirements for each kind of token. Hence, it is essential to understand which of these requirements can be availed by the organization's employees as well as consumers before it is implemented.

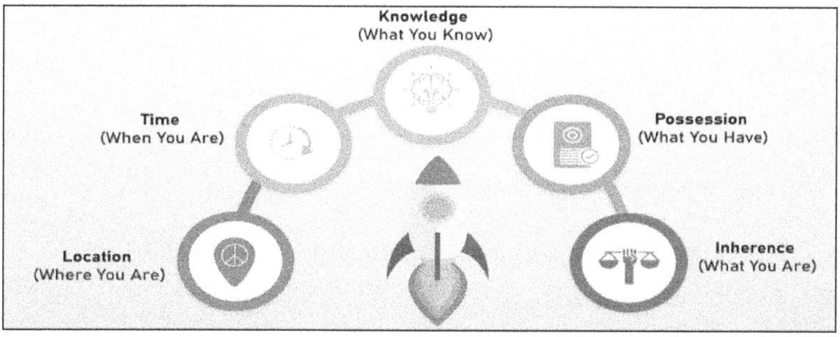

Figure 4.12 *Key Components of Multi-Factor Authentication*

- **Knowledge (What You Know):** A password, a security question, or a PIN that ideally only the user knows is called the knowledge. This is usually the most widely used and the first level of authentication.
- **Possession (What You Have):** This is an authentication based on something that the user has, for example, a mobile phone, a SIM card, a smart card or a key fob. For this reason, if a hacker is able to access the password, they would need to obtain one of these possessions in order to successfully penetrate the system.
- **Inherence (What You Are):** This authentication is based on unique biological traits such as fingerprints, the iris of the eye, and facial features. This typically requires reader hardware, a database, and software to process for authentication.
- **Location (Where You Are):** This is the address where the user submitted their access request. It makes use of the request's IP address and, if accessible, the user's geolocation.
- **Time (When You Are):** The time of the user's access request determines this. For instance, the request is rejected if the worker's work hours are from 9 a.m. to 5 p.m. and they haven't been given the ability to log in after that time.

Types of Multi-Factor Authentication (MFA) and Biometric Authentication

The following table lists the many forms of biometric authentication and multi-factor authentication (MFA) techniques:

Table 4.2: Types of Multi-Factor Authentication (MFA) and Biometric Authentication.

Aspect	Multi-Factor Authentication (MFA)	Biometric Authentication
Definition	a security technique that uses three or more verification factors: your identity (biometrics), your knowledge (password), and your possessions (token, OTP).	a security technique that uses distinctive behavioural or physical traits, such voice patterns, fingerprints, or facial recognition, to confirm identity.
Types of Factors	Combines multiple factors: knowledge (passwords, PINs), possession (smartcards, OTPs), and inherence (biometrics).	Uses one factor: something inherent, such as fingerprints, face recognition, iris scan, or voice.
Security Level	Provides multi-layered security by requiring more than one verification method, making it harder for attackers to bypass.	High security based on unique physical traits, but can be vulnerable to spoofing or errors.
User Experience	Requires users to provide more than one authentication factor, which may involve extra steps (e.g., entering a password and then verifying with an OTP).	Provides a seamless, fast user experience, often requiring only a fingerprint scan or facial recognition.

Aspect	Multi-Factor Authentication (MFA)	Biometric Authentication
Implementation Complexity	More complex to set up as it requires integrating multiple authentication methods, including hardware tokens, OTP generation, etc.	Easier to implement for specific use cases (e.g., fingerprint scanning), but may require sophisticated sensors and software for accuracy.
Examples	Entering a password + receiving an OTP Using a smartcard + PIN	Fingerprint recognition Facial recognition Voice recognition

Source: (Self-generated)

Multi factor authentication (MFA) is the practice of requiring multiple verification factors in order to gain access to something, such as passwords, tokens, or biometrics, to make unauthorized access more difficult but potentially more complicated for users. On the other hand, when we talk about Biometric Authentication, it is based on some unique physical traits like the fingerprints, facial recognition or voice pattern, and provides a seamless and fast experience but also it has a vulnerability to the spoofing or sensor limitations. MFA offers robust and multi layered protection while biometric authentication prioritizes convenience, so both offer much of what the other lacks, based on the need for security.

4.5 The Future of AI in Fraud Detection and Financial Markets

In an increasingly sophisticated financial fraud, AI continues to become more vital in detecting and in stopping such fraudulent activities. Real time advanced machine learning models take in huge shifts of transactional data and monitor their anomalies for possibly fraudulent occurrences. Unlike traditional rule-based systems, because AI driven fraud detection is a continuous process of learning from the new patterns, it is more effective in identifying not only those already known threats, but even the new threats that arise in the system. AI takes full advantage of behavioral analytics, anomaly detection and predict modeling to limit cyberattacks and financial risks for financial institutions.

While AI has a known use in fraud detection, AI is completely changing financial markets by enhancing risk assessment, algorithmic trading, and customer service. With AI powered predictive analytics, more accurate market trends can be evaluated, any credit risks can be assessed better and investment strategies can be optimized better than what the traditional models can do. AI is being used as a source of power and fuel for high frequency trading algorithms, which harvest market data at such fast speeds as to make profitable opportunities and execute trades within milliseconds. Also, AI fuelled chatbots and virtual assistants are assisting the client encounters, furnishing with individual financial guidance and robbery advance alerts in genuine time.

In the foreseeable future, the integration of AI with the blockchain and quantum computer could possibly build on the revolution in the fraud detection and security in the finance field. Blockchain technology ramps up on improving transaction transparency and lower the risk of tampering while quantum computing breaks the existing cryptographical methods that can reshape the standard of cybersecurity. AI systems will grow progressively more complicated and thus begin regulating and the need of regulatory framework will be there to regulate the deployment of AI and protect the Entity of financial decision making to be on a firm basis of ethical AI. There will be a subtle balance of innovation, security and compliance of the future of AI in fraud detection and financial markets.

4.5.1 AI's role in securing digital payments and cryptocurrencies

Digital and decentralized financial systems have spread and therefore cryptocurrencies and digital payments' security have a primary concern. Fraud prevention measures are greatly strengthened by AI which makes certain that transactions are authentic and prevents financial risks. Using AI powered technologies, the financial institutions, payment processor also the cryptocurrency network can detect and curtail the fraudulent activities in real time, thus securing the use of this services by the business and the users.

Digital payments can be more accurately fraud detected and risk managed by AI in analyzing transaction patterns to identify the anomaly and flag the suspicious activities. The parameters such as the location, device type etc. help machine learning algorithms to detect deviations of unusual spending, fake accounts and unauthorized transactions. AI driven behavioral analytics on top of it further helps in security by monitoring the user's interaction behavior and trigger alert in case deviation from the normal behavior, requiring more steps of additional authentication such as Multi-Factor Authentication (MFA).

AI serves the security in the cryptocurrency ecosystem through the blockchain integrity strengthening, detection and prevention of such fraudulent activities, and also from those cyber threats. Anomaly detection using AI can detect the doubling of spend, gain unauthorized access to the system, or other potential ways to manipulate the market, whilst smart contract auditing codes to ensure the security code and decrease in the vulnerability of that code. Apart from improved security, AI has also helped in making cryptocurrency wallet secure by using behavioral biometrics (such as typing speed & device holding patterns) that stop unauthorized access and cyberattacks in cryptocurrencies wallet.

For the purposes of looking ahead, AI is expected to incorporate technologies like blockchain, quantum computing, adaptive authentication systems and others. Financial security will also be fortified by AI driven biometric verification, risk base authentication and ant money laundering compliance monitoring. But as AI matures, so do these frameworks, for instances to be transparent, to employ ethical AI and to keep data privacy. Naturally, balance in balance between marketable innovation and compliance with security is the future of AI's role in fraud detection and financial markets (Agarwal et al., 2020).

4.5.2 Predictive analytics for evolving fraud tactics

Detecting and preventing fraud has a wide range from traditional financial systems to deviating technology through cryptocurrencies and digital payments. As fraud schemes metastasize, financial institutions, amongst others, in the corporate payments space, payment providers and others are seeking advanced AI and machine learning algorithms to fight back against growing fraud tactics. The prediction analytics exploit historical information, statistical algorithms, and machine learning models to make forecasts of future fraudulent behaviors, discover trends, and counteract such fraud before it happens.

Understanding the importance of predictive analytics in fraud prevention

Fraud prevention is made possible by Predictive Analytics as it forecasts potential fraudulent activities before actually taking place by analyzing the historical data. Predictive models based on AI enable organizations to anticipate potential fraud patterns and determine the levels of risk, as well as taking preventive steps to combat against possible threats. Through processing huge amount of transactional data, these models can discover such anomalies and correlations as unusual transaction locations, improper transaction amounts, or unfamiliar devices usage. By gaining these insights, financial institutions, e-commerce platforms and many other businesses can act early before fraud occurs, and therefore minimize financial losses and protect the customers' trust. Further, predictive analytics not only sharpens the security but also assists in reducing the manual efforts involved in fraud detection and thereby enhances the operation of the fraud prevention teams.

As the fraudsters find ways to bypass the old methods, it becomes ineffective to rely on these traditional rules of blacklisting, rigid transaction thresholds, etc. In particular, many cybercriminals have a proclivity toward developing new tactics upon which to operate like synthetic identity fraud or extremely sophisticated phishing attacks making the static rule-based approaches less effective over time. However, predictive analytics is meant to be dynamic, always learning and detecting the slightest shifts in fraudulent behavior from new data. The predictive models get more accurate by using the more advanced machine learning algorithms to make differentiations between a legitimate transaction and a fraudulent activity. The outcome is not only to help to improve a previous version of engaging fraud in the way that it hasn't been seen before, but also to help minimize the number of false positives so that customers are not wrongly flagged as fraudsters, which in turn creates a better customer experience.

Key Techniques in Predictive Analytics for Fraud Detection

Advanced techniques are used to predict and fight evolving techniques of fraud by analyzing large datasets with the help of machine learning and detecting the anomalies. Complex fraud pattern contained in massive transaction record is recognized through clustering algorithms which enable the models trained based on historical fraud data to predict future risk and make continuous improvements in accuracy. By continually solving for daily problems within the company, NLP improves fraud detection by looking to emails, chats, and online interactions for phishing scams and social engineering attempts to stop fraud before any real harm occurs. Furthermore, time series analysis helps detect unusual transaction behaviours like sudden upsurge in activity or odd hours of activity which is typically a fraudulent attempt. And further in, anomaly detection further strengthens security through the creation of behavioral baselines for users and accounts and their predictive model that will alert when users and accounts deviate from normal behavior. It is particularly helpful to assist organizations in identifying new fraud techniques that do not fit within the existing pattern, so that they can continue to be proactive about their fraud prevention efforts (BluEntTech, 2023).

Predictive Analytics and Evolving Fraud Tactics

Constantly being sought out by fraudsters, financial systems must be vulnerable to an evolving set of fraud tactics and the only way to stymy fraud is through predictive analytics. The biggest threat here is synthetic identity fraud, i.e. where the criminals come up with fake identities with a mix of real and fictitious information and open accounts and make transactions using them. By detecting such fraud through identifying inconsistencies of personal data, mismatch of identity attributes, and uncommon application behaviors, predictive analytics does this. The second most common scheme is account takeover fraud where these people are able to get access to real accounts and use it to conduct fraudulent transactions. Of course, the underlying predictive models can be used to identify the suspicious activities as multiple failed login attempts, sudden changes in devices or locations, or modifications in account details. Fraudsters lure victims into divulging private information in phishing and social engineering attacks as well. Predictive analytics review email, SMS, social media communication to detect suspicious language and behavioral patterns which are indicating phishing attempts. On top of that, cryptocurrency fraud such as pump-and-dump schemes or ICO scams is also being rampaged. By removing the unpleasant surprises from trading, predictive analytics can identify unplayable trading behavior, big scale market manipulation and abnormal transaction patterns of cryptocurrency exchanges, financial institution can avoid the unpleasant surprises and prevent their systems from being exploited. (Nikhil Saini, 2025).

Multiple Choice Questions (MCQs)

1. What is one of the primary roles of AI in financial security?

A. Reducing human involvement in banking operations

B. Enhancing transaction speed only

C. Detecting and preventing fraudulent activities

D. Replacing all financial analysts

2. How does AI enable real-time fraud detection in financial transactions?

A. By manually reviewing transactions

B. By applying machine learning algorithms to identify suspicious patterns

C. By blocking all transactions above a certain amount

D. By only allowing transactions from verified accounts

3. What is AI-based risk scoring used for in financial transactions?

A. Assigning a fraud probability score to each transaction

B. Replacing manual customer verification

C. Predicting future financial trends

D. Blocking transactions based on location alone

4. How does AI help in detecting money laundering activities?

A. By analyzing historical transaction patterns and identifying anomalies

B. By limiting transaction amounts for all users

C. By manually reviewing customer transactions

D. By only monitoring high-value transactions

5. **What is one key benefit of automating AML reporting with AI?**
    A. Reducing the need for regulatory oversight
    B. Eliminating the requirement for compliance teams
    C. Speeding up the identification of suspicious financial activities
    D. Preventing customers from making international transactions

6. **How does AI-driven behavioral analytics improve fraud prevention?**
    A. By tracking customer transaction habits and identifying unusual deviations
    B. By blocking all new transactions after business hours
    C. By requiring customers to change passwords frequently
    D. By ensuring all transactions are processed manually

7. **What is the purpose of AI-powered biometric authentication in financial security?**
    A. To replace passwords with unique physical or behavioral traits
    B. To eliminate the need for financial regulations
    C. To restrict account access to business hours
    D. To allow only face-scanning authentication

8. **How does AI contribute to securing digital payments and cryptocurrencies?**
    A. By blocking all cryptocurrency transactions
    B. By monitoring blockchain transactions for anomalies
    C. By eliminating the need for security in digital payments
    D. By manually verifying every transaction

9. What is a key advantage of predictive analytics in fraud detection?

A. It prevents fraud before it occurs by identifying risk patterns

B. It only works after fraud has been reported

C. It replaces human fraud analysts entirely

D. It limits the use of online banking services

10. How do AI-powered fraud detection models stay effective against evolving fraud tactics?

A. By continuously learning from new fraud patterns and adapting

B. By following only pre-set fraud rules

C. By blocking transactions based on random selection

D. By requiring customers to verify their identity daily

Answers

1	2	3	4	5	6	7	8	9	10
C	B	A	A	C	A	A	B	A	A

Chapter 05

Future-Proofing Industries with AI

5.1 Introduction to AI in Future-Proofing Enterprises

In a rapidly digitalized and interdependent world, industries are evolving, organizations need to implement and leverage advanced technologies in order to survive, be resilient and be responsive. Artificial intelligence has become a game changer in guaranteeing enterprises 'future proof ness through uncertainty handling, site operations optimization, and a never-ending innovation. AI enables strategic decision-making process automations, identifies potential risks which may escalate, and keeps the organizations ahead in every aspect of the market. Using AI, companies can not only increase efficiency but also get a huge competitive edge over their competitors in the industry (Schildt, 2020).

One of the real key contributions of AI that FW – enterprises need to future proof is precisely its prediction capability. Machine learning models use historical and real time data to make market forecasts, customer behavior, financial risk and operational inefficiency. The ability to be able to make out this foresight makes it possible for the business to proactively readjust their strategy, allocate their resources more effectively and mitigate any eventualities before same disrupts the way they operate. Also, AI driven automation minimizes manual workflows, eliminating error rate, increasing productivity and focusing human workers on high value items that advance innovation and the long run development. Integrating AI into decision making process will help the organizations to react swiftly to change in market fluctuations, changing consumer demands.

In addition to operational efficiency, Enterprise Resilience is strengthened by AI by improving cybersecurity, increasing regulatory compliance, increasing risk management and optimizing supply chain. It is a continuous monitoring of network activity, threat detection, vulnerability identification and real time cyber threat response that is powered by AI, to protect critical business infrastructure from the possible breaches. It is also the same with AI driven compliance tools that help organizations work with complicated regulatory environment and remain legal, with minimal financial and reputational risk. AI optimizes logistics; answers every question on demand forecasting; and facilitates seamless coordination between global operations to cut down on costs and improve the efficiency of delivery. Integrating AI into core business functions not only allows enterprises to build a solid base for long term sustainability, agility and success but also in an increasingly volatile and technology driven world, AI will facilitate them to operate in a smooth manner.

5.2 Emerging AI Trends in Security, ERP, and Risk Management

The modern security, ERP and risk management fields are experiencing a rapid evolution with Artificial Intelligence, enabling organizations to gain access to high-powered tools to make operations more efficient, more secure and reduce risks in business. With the advancement of the AI technologies, enterprises are using the machine learning, natural language processing (NLP), deep learning and automation to detect threats proactively; to improve enterprise workflows and make better decision processes in real time. The complex and increasingly interconnected business environment has led to AI driven solutions for businesses to redefining how resources are managed, digital assets are secured, world is scanned to understand new markets to enter and regulatory compliance is managed. Adopting AI not only improves productivity, but it also helps organizations advance themselves from the emerging risks and dynamic market conditions (Access et al., 2024).

AI powered systems are changing how security detects threats, and the ways to prevent and resilience to security attacks. Real time data analysis through advanced machine learning algorithms are used to analyze vast amount of data, looking for unusual pattern, a suspicious behavior or anomaly that might indicate a cyber threat, data breach or a fraudulent activity. With its roots in predictive analytics and AI powered, security solutions powered by AI will in turn enable businesses to think and act proactively to protect the sensitive information or vulnerability of a digital infrastructure. Secondly, AI powered identity verification, biometric authentication and behavioral analytics help to enhance access control, limit risks of unauthorized employee access to critical business systems and restrict authorized personnel from interaction with the critical business systems only. Automated incident response is also happening with the help of AI, wherein, cybersecurity teams are capable of detecting threats, investigating and neutralizing them in real time on a way faster manner thus causing the impact of cyberattacks minimized.

AI is automating and providing intelligent insights to enable a better strategic decision making, operational stability and business continuity, in ERP and risk management. AI based ERP solutions make resource planning smooth, work idealization and eliminates unproductiveness, along with human mistakes. Data analytics that are driven by AI helps businesses predict the demand, optimize the inventory, make better of supply chain management and lay out excellent financial planning. An additional benefit of AI powered chatbots and virtual assistants in ERP systems is that they improve user experience in real time, answering the routine questions automated. At the same time, in risk management where AI is being applied to assess financial, operational and compliance risks with greater accuracy and speed. Big data analytics with the help of AI powered risk modeling is used to predict vulnerabilities, disruptively and strategies to avoid loss of money by using proactive strategy. Integration of AI into security, ERP and risk management, allows organizations to build a strong digital ecosystem that helps strengthen the organizations' agility, resilience and long-term success in this new and competitive world of interactions and marketplace.

5.2.1 AI advancements in cybersecurity and enterprise automation

It is time to look at cybersecurity and enterprise automation through Artificial Intelligence (AI) lens and to recognize that artificial intelligence is revolving the modern working system to boost security measures, gain efficiency and automate enterprise in their overall process. As the landscape of cyber threats and business processes become more complicated, artificial intelligence (AI)-driven solutions can offer predictive insights, real time threat detection and intelligent automation to address risks and streamline workflows. The AI not only dimities the need for human intervention, but for the purposes of cybersecurity defenses as well as enterprise functions, it makes things more accurate, more reliable, and faster. Since it is rates up across industries, digital transformation is huger Affibody as a fixture of business assets protection and operational agility (Jagreet Kaur, 2019).

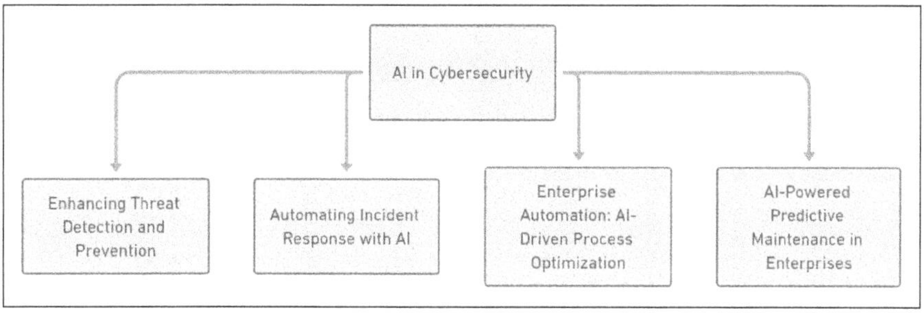

Figure 5.1 *AI in Cybersecurity: Key Applications and Use Cases.*

Source: (Self-Generated)

1. AI in Cybersecurity: Enhancing Threat Detection and Prevention

The cybersecurity threats have advanced tremendously and they demand advanced AI-driven solutions to ward off any breach. Machine learning based on AI is used in the design and development of programming algorithms that can analyze a large volume of data in real time and identify patterns of behavior that can indicate a cyber threat in seconds. These systems are proactive and continuously stop malware attacks, phishing attempts as well as unauthorized access by predicting and neutralizing a threat before it escalates further. Also, by increasing the identity verification with biometric authentication, behavioral analysis and anomaly detection, only authorized persons may access the sensitive information.

AI is also being used in network security monitoring where it is running a scanner and an analyzer both simultaneously against the network traffic to detect any suspicious activities. Unlike conventional rule-based security systems, it is possible for AI based system to learn from new attack patterns to help prevent zero-day attacks. AI integrated with SIEM platforms integrate AI with the security information and event management systems (SIEM) and the organizations use it to automatically find out the threat, correlate security events, and provide meaningful insights in real time.

2. Automating Incident Response with AI

Incident response is being changed by automation of threat detection, investigation, and mitigation due to AI. Security orchestration and automated response (SOAR) using AI platform helps security teams with efficient threat handling services by generating automatic alerts, suggesting countermeasures and automating predefined attacks to prevent the attacks. These platforms link with different cyber security tools to enable businesses to have a centralized security management system and respond to cyber threats fast and well.

Also, AI based powered chatbots and virtual assistants are being deployed to the aid the IT and security teams in securing IP address range, detecting security issues, recommending solutions and streamlining response process. Using AI in security can reduce the time it takes for handling security incidents by a business by a great deal, can keep you from losing money in all, and will make your security posture as a whole stronger. AI is crucial in modern cybersecurity frameworks due to its ability to analyze huge number of security log, discover attack patterns, and create automated response.

3. Enterprise Automation: AI-Driven Process Optimization

In addition to cybersecurity, AI has brought in significant automation for the enterprise field, enabling efficiency at different levels with automation. Robotic process automation (RPA) can be driven by AI, therefore automation takes place in repetitive tasks such as data entry, invoice processing and customer service inquiries to free employees to engage in more rewarding areas. The enterprise resource planning (ERP) systems are integrated with Artificial Intelligence to boost the workflow process by optimizing it, also the inventory management while forecasting the demand and making a better decision based on their data. By predicting demand fluctuations, the automation tools based on AI make supply chain management more efficient as well as optimizing logistics as well as enabling seamless coordination across the worldwide operations.

Similarly, human resource management (HRM) is also being automated by AI in the forms of an automated recruitment, employee performance analysis, and workforce optimization. AI chatbot helps HR team to answer employee queries, book interviews as well as labor market trends analysis. Moreover, AI-enabled business intelligence solutions allow organizations to achieve greater understanding of operational data, of which executives can come up with informed strategies.

4. AI-Powered Predictive Maintenance in Enterprises

These strategies have also been extended further by AI to enhance predictive maintenance practices that allow enterprises to minimize downtime and increase equipment's reliability. A combination of AI algorithms analyzes sensor data from IoT devices as well as industrial machinery to predict when machinery or IT infrastructure is going to fail so that companies can react before they disrupt. Through this approach, operational efficiency is improved, the costs of maintenance are minimized, workplace safety is increased, and the life time of the critical assets are stretched. Specifically, AI powered predictive maintenance is great for manufacturing, energy and transportation industries due to the significant financial and operational losses that occur from unplanned equipment failures.

5.2.2 AI's evolving role in risk-aware business strategies

Artificial intelligence (AI) brings it about, through advanced analytics, real time monitoring and prediction, the way businesses work through risk management. Historical data, human expertise and static risk models were traditionally used by companies to assess and ameliorate risks. Nevertheless, these are traditional methods that often do not meet current business needs in a fast-changing environment where new risks arise frequently. With AI businesses are able to switch from reactive to proactive risk management by at all times analyzing large amounts of structured and unstructured data. Using machine learning

(ML) and natural language processing (NLP), AI helps the organizations to discover the patterns, anomalies, and correlation points which may indicate the potential threats and then organizations make the decision capable of minimizing the disruptions by making the data driven decision (Briefs & Business, 2018).

The most significant benefit of AI in risk-aware business strategies is its capability in predictive risk assessment. Unlike this traditional method which depends solely on past incidents to predict future risks, AI is able to analyze real time information from all kinds of channels (financial markets, supply chain, cybersecurity networks, regulatory databases) and not only for the case. For instance, AI models analyze unusual transaction patterns in order to identify fraudulent activities in the financial sector, while in healthcare, AI predicts patient risks using patient histories as well as live health indicators. A similar example in manufacturing and supply chain management is AI predicting potential disruptions caused by equipment failures, weather condition, or the geopolitical instability to develop the contingency plans for respective companies. Using AI driven predictive analytics a business can predict risks before they develop into major losses in financial terms, stand down operations and other costs which arise due to issues.

Another very important contributing area to AI transformation in risk management is cybersecurity. As cyber threats get more sophisticated cyberagents require that businesses have intelligent security systems that could detect and responds to attacks in real time. AI based security solutions constantly monitor the network activity, look for anomalies and find out possible security breaches before it causes major damage. These are the systems which uses the machine learning algorithms to find the patterns that reveal the cyber threat such as the phishing attempts, malware infections and unauthorized access. AI powered fraud detection system also enables the security of transaction by analyzing spending behavior and flag suspicious activities. Integrating artificial intelligence in cybersecurity framework can strengthen organizations' immunity to cyber threat and protect sensitive information from get breached.

It is also revolutionizing compliance and regulatory risk management by automating compliance tracking and making sure that it complies to the industry standards. But regulations continue to change dynamically, which can be a hard job assuming that businesses will keep sufficient up with the compliance obligations. Frustrated by the lack of machine-readable descriptions in existing legislation and regulatory documents, the compliance tools that rely on AI power seek to decipher legal text using NLP in order to sass out distillable information that indicates something needs to be done to satisfy the letter of the law. In the industries of finance and healthcare particularly, AI driven compliance monitoring systems track transaction and operation for non-compliance risks detection, which reduces the chances of regulatory penalties. It also helps with automated reporting, which relieves the compliance teams from the burden but at the same time, ensures accuracy and efficiency in regulatory filings. Combining AI with compliance strategies helps businesses cut down on regulatory processes, eliminate risks and make processes more transparent by integrating it.

A second important application of AI in risk management is its aid to decision making. The decision support systems pushed by AI aggregate and analyze the mountain of data at real time and takes the burden of interpreting it from executive hands, helping them determine how to assess the risks and put together mitigation strategies. AI lets organizations test out risk scenarios and scenario analyses to evaluate the risk and identify the best reaction. AI models can be used for supplying supply chain management, which can simulate disruptions and then recommend other possible routes or suppliers to keep the operations going. AI driven simulations aid investors to determine the market risks and make wise investment decisions in financial risk management. Using AI for strategic decision making helps businesses become more agile, lowers uncertainty, as this approach allows businesses to achieve long term stability.

AI has also improved crisis management as well as business continuity planning. Organizations can face the issues such as natural disasters, economic downturn, cyber-attack and operational failures. They help businesses respond more effectively in emergencies through real time situational awareness about the situation within. So, tools that respond in the disaster response area using AI as a trigger are powered by the data extracted from weather forecasts, social media, sensor networks and such to predict disasters, and suggests taking proactive measures. In the same way, AI chatbots and virtual assistants help in crisis communication by keeping the employees and the stakeholders updated and following the emergency protocol at hand. AI in supply chain management helps businesses stay afloat by spotting vulnerabilities and providing the ways to escape them in case of disruptions. It gives organizations the opportunity to use the AI on crisis management to decrease downtime and financial loss, and enhance overall resilience.

5.3 AI Integration with Blockchain and IoT for Enhanced Security

The security is being totally reformed via the use of AI, blockchain and IoT to create more responsive, intelligent and resilient infrastructure. AI is able to continuously analyze the data in real time to perform threat detection and anomaly identification in real time monitoring of IoT devices. Detecting irregular patterns, predicting the possible cyber threats and reacting proactively can help minimize the risk of security breach, such systems can be used. Blockchain ensures the data integrity by tamper proof records, and decentralized authentication mechanism that is immune to unauthorized access and manipulation of the data. AI driven analytics can be integrated with blockchain's immutable ledger to build an extremely secured framework with which organizations can protect IoT ecosystem from all cyber threats, data breach and malicious intrusion (Anamika Pandey, Balamurugan Balusamy, 2023).

The IoT generated data is continuously scanned through the use of AI powered machine learning models which find out security risks, predicts vulnerabilities, even before cyber-attacks happen. Also, these models improve the work done on fraud detection, automate the security response, and optimize network performance by real time threat identification & remediation. The combination of blockchain technology allows identity management to be more decentralized, deals with encrypted transactions and transparent and secure data exchanges. The enforcement mechanisms in smart contracts are further automated, and if needed, ensure compliance and allow people to trust interactions between connected devices without humans' involvement.

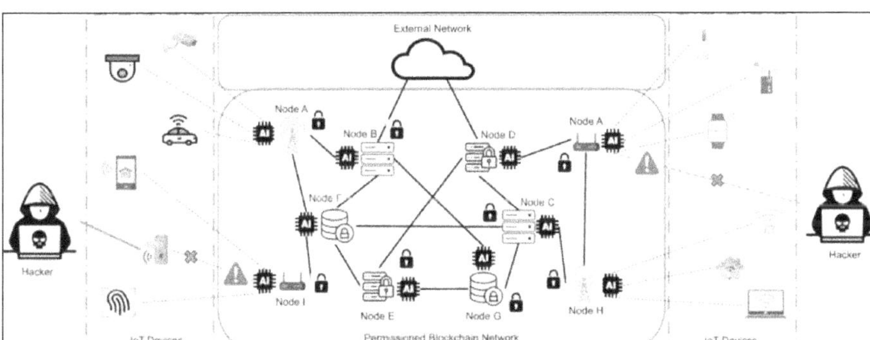

Figure 5.2 *IoT Security with Blockchain.*

Source: (Dhieb et al., 2020)

Blockchain, AI, IoT synergy is used in the fields of healthcare, finance, smart cities, supply chain management and industrial automation in the context of cybersecurity. It protects patient sensitive data and also provides secure medical device communication for healthcare. It is used to secure transaction and prevent fraud in finance. In smart cities it protects the cyber threat on traffic management systems as well as energy grids. These were technologies that were evolving, and based on that, it will be possible for businesses and governments to create more robust security frameworks for privacy of data, regulatory compliance and uninterrupted process. Integration of AI, blockchain and IoT enables industries to build a secure ecosystem that is not only effective in solving current challenges but more than that, it provides an insight into the future threats and predicts.

5.3.1 AI and blockchain synergy for data security and transparency

AI and blockchain combination are disrupting the data security and transparency as it joins the advanced intelligence and an imperishable and purely decentralized approach. AI helps in improving cybersecurity by examining huge amount of data at real time as it decides the anomalies, and predicts the potential threats before they turn into a security breach. This makes it possible to automate monitoring, recognition of patterns, and decision making, so that businesses can always lead the charge against the cyber threats. Meanwhile, blockchain guarantees the data integrity because every transaction is recorded in a secure, transparent, tamper proof manner. This combination helps reduce the risk of data manipulation, unauthorized access and attacks by cybers, making it a requirement for organizations that deal with the sensitive information (Von Braun et al., 2021).

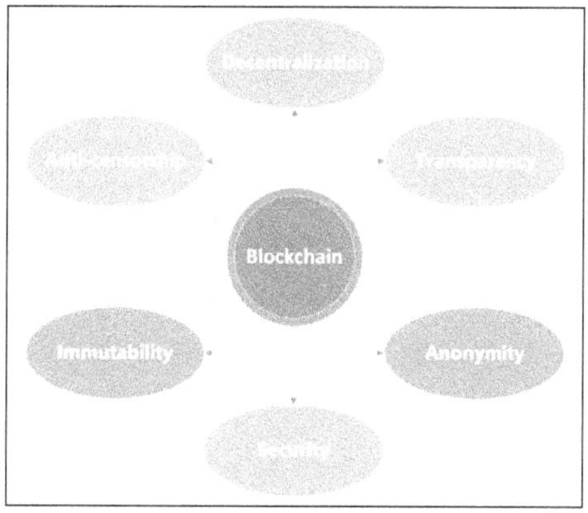

Figure 5.3 *Key characteristics of blockchain.*

Source: (Bhumichai et al., 2024)

The key advantage of integrating AI with blockchain is that its use enhances the transparency achieved in data management. Blockchain's technology based on distributed ledger ensures a given transaction is securely recorded, verifiable and available to the authorized stakeholders without fear of data falsification or unauthorized modifications. AI adds transparency by automating a compliance checks, analyzing discrepancy and enforcing regulatory standard. It is of particular use in the domain of finance, healthcare, supply chain management and government operations, where simply keeping track of what has happened is the order of the day. Using AI, an organization can analyze blockchain data for the detection of fraud, regulatory violations, or suspicious transactions in order to protect business interests and at the same time comply with ethical, legal and other kinds of requirements, thus reducing the possibility of human error.

Apart from security and transparencies, integration of AI and Blockchain produces operational efficiency and innovation. It is the use of AI-powered smart contracts in the blockchain network that streamlines business processes, reduce operational costs and eliminates the need of intermediary to automate secure and self-executing agreements. Resource allocation can also be optimized and inefficiencies can be detected with AI driven insights along with better decision making and security of stored, and transmitted data. For example, in the supply chain logistics sector, AI can monitor product authenticity, and blockchain can create an unalterable record of every stage of the supply chain, this synergy is particularly transformational. The combination of evolving AI and blockchain technologies only increase the need for such an indispensable pillar of digital trust, secure transactions, verifiable records, and a strong cybersecurity base of support for businesses worldwide.

5.3.2 AI-powered IoT risk management solutions

The Artificial Intelligence (AI) coupled with the Internet of Things (IoT) has revolutionized risk management in different industries. Real time data on IoT devices is another beast altogether, especially when AI is in the equation to crunch it, to secure the IoT device and enhance its functionality. By using machine learning algorithms, predictive analytics, and automation, an AI powered IoT risk management solutions have the ability to detect vulnerabilities, prevent cyber threat and system's resilience. More so, these solutions are necessary in highly sensitive and data security sensitive sectors like healthcare, manufacturing, finance, and smart cities (Rieple et al., 2023).

By leveraging AI, the IoT systems are capable of adapting dynamically to any changes in the environment by a pattern analysis, anomaly detection, and e automatic threat response. And traditional risk management types tend to fall short when it comes to complexity and scale of IoT networks, necessitating that AI be a pipeline must for proactive monitoring and risk mitigation. AI will make IoT ecosystems more secure and reliable whether it's in securing industrial control systems, preventing data breach in connected healthcare devices or improving the supply chain operation. Some key subheadings of how AI is changing the game of IoT Risk Management are listed below:

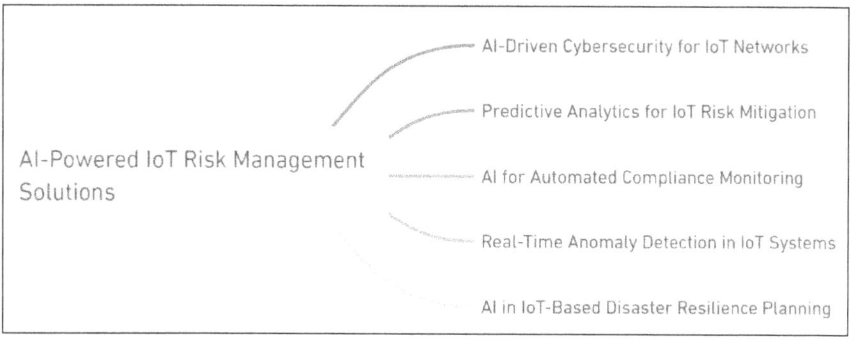

Figure 5.4 *AI-Powered IoT Risk Management Solutions.*

Source: (Self-Generated)

1. AI-Driven Cybersecurity for IoT Networks

The more connected the IoT devices become, the larger attack surface against which cybercriminals have to work. Through its AI-driven tweaking, cybersecurity solutions look at massive amounts of the data coming from the IoT networks in real time, and detect anomalous behavior that could be indicative of cyber threats. With the help of machine learning models, AI can disrupt and predict malware attacks, unauthorized access, and data breaches since it can predict cases. They convert the concepts into automated threat response mechanisms again, AI-based firewalls and intrusion detection systems to deal with the threats before they can deal a lot of damage. Endpoint security is also aided by AI, since all IoT devices have to adhere to strict security protocols and thus minimize vulnerabilities.

2. Predictive Analytics for IoT Risk Mitigation

AI based predictive analytics help organizations to predict the future risks that are going to occur in future. An example of detecting patterns and predicting failures in industrial equipment, transportation systems and healthcare devices would be by analyzing historical and real-time IoT data. This proactive approach provides a business an opportunity to prevent problems, enjoy less downtime and experience fewer operational disruptions. Predictive analytics work in sectors like smart grid and autonomous vehicles, so as to keep system reliable by detection of irregularities and recommending maintenance or corrective actions. The use of AI powered insights allows organizations to take data driven decisions that will reduce the risk and optimize the system efficiency.

3. AI for Automated Compliance Monitoring

A big issue in such industries where IoT technology is used is the regulatory compliance. Such compliance monitoring systems are AI powered and scan IoT generated data to confirm that the industry regulations as well as legal standards are being met. Not only these systems can automatically detect compliance violations, produce real time alerts, and generate detailed audit report. Compliance with the regulations becomes easier with AI that keeps track of IoT environments to detect any inconsistency, lowering the risk of being penalized in terms of data breaches and legal penalty. Along with this, AI assists companies in meeting emergent regulatory demands by discovering legal documents for analysis and update the compliances framework.

4. Real-Time Anomaly Detection in IoT Systems

AI in IoT risk management has one of the key advantages of being able to detect anomalies in real time. Anomaly detection models are run on the IoT data streams using AI and continually monitor what is happening in the streams for any deviations from normal behavior patterns. In fields like industrial automation, smart homes, and connected healthcare devices, this is more so important to remove undetected anomalies as it could result in equipment malfunction, safety hazards or security breaks. The combined use of such anomaly detection tool powered by AI, leads to minimizing false alarms because it narrows the difference between normal variations and risks so that genuine risks receive timely and accurate responses.

5. AI in IoT-Based Disaster Resilience Planning

In disaster management, IoT devices are very important to collect data from environmental sensors, weather monitoring systems, and infrastructure networks. Disaster resilience is improved by the fact that AI uses this data to predict and respond to natural disasters like earthquakes, floods and fires. The IoT systems powered by AI can give early warnings, make emergency response efforts better, and can allocate resources in a right manner. AI powered IoT networks in smart cities help in real time traffic, power and safety measures for public in case of a crisis. The combination of AI with IoT enables government and organization to become more prepared for disasters and reduce losses.

5.4 AI-Driven Governance, Risk, and Compliance (GRC) Strategies

Business operations are also governed, risked, and complied with in Governance, Risk, and Compliance (GRC). Given the ways and intricacies of an infinitely changing and intricate regulatory environment, traditional GRC models, characterized by non-computable processes, static policies and reactive procedures are proving to be inadequate. Due to its automated, data driven, and predictive capabilities, Artificial Intelligence (AI) helps transformed the GRC frameworks and comply fully, giving the risk assessment better and making it easy to make governance decisions. GRC strategies that use AI for supporting the implementation use machine learning, NLP and analytic computing to develop an active, city, streamlined and scalable system to monitor, manage and manage the challenges brought on by regulatory concerns as well as those threats to the organization (By, Anthony TAEditedrantino, 2008).

It has repositioned GRC by allowing the automation of processes that in the past have been manual and prone to errors by humans. Risk identification and assessment one of the key applications of AI in GRC. From business operations data, financial reports, regulatory changes, they monitor and analyze structured and unstructured data of business operations, and detect potential risks with the help of machine learning algorithms. Organizations can intervene proactively when risks escalate to identify different patterns and patterns of threats, these systems can identify them. Furthermore, continuous updates from real time data in AI models form the basis for risk assessments that remain dynamic and spawning provision of risk management strategy that is dynamic in nature.

In another area, AI has profoundly influence regulatory compliance. Regulatory Changes through regulation can be monitored with the help of the AI powered toil which can keep a check on legal aspects of respective businesses and make sure they meet the changing needs. With NLP, AI is able to scan large amounts of legal documents to determine the compliance obligations, and to flag potential violations. An automated compliance monitoring will minimize human errors and ease the process of getting fulfilled onto the reporting requirement of organizations. Also, there are AI driven chatbots that can help the compliance teams with the real time guidance, query related policy and generate the compliance reports as per the time demand.

AI has been a game changer, and its merit in the field of cybersecurity and fraud detection has found no comparison. The security frameworks can be powered by AI to analyze vast amount of network traffic and user behavioral data that mark out the anomalies indicating possible threats or any kind of fraudulent activities. Such systems will detect potential security breaches, unauthorized access attempts and suspicious transactions where organizations can respond quickly enough to mitigate risks. Their AI powered cybersecurity solutions also help the companies in complying to variety of industry standards inscribed as ISO 27001, NIST and GDPR, by following the security measures as per the regulatory expectations.

In addition to that, AI also has an important place in corporate governance by enabling data-based decision making. AI powered analytics are used for governance framework in terms of providing insights on these financial risks and operational inefficiencies and ethical considerations for business policy making. AI driven cognitive systems enables executives to assess the level of compliance risk, and evaluate various governance scenarios, taking the decisions which are aligned to the regulatory requirements as well as company's goals. In addition, AI can also be used for scenario analysis on crisis management, in order to predict types of challenges that may arise in case of a crisis and develop strategies to handle such situations.

The advantages of AI driven GRC strategies over the traditional ones are many, thus governance, risk management and compliance is more efficient and effective. One of the first advantages is that it makes the operation more efficient. AI automates repetitive task such as compliance monitoring, risk assessment and fraud detection and that reduces the GRC professionals manually workload. It is this, that allows for organizations to effectively allocate resource and emphasis on strategic decision making instead of administrative tasks.

The second benefit is in real time risk insights. AI continually assesses internal and external data sources and means of delivering up to date risk assessments. It provides real time monitoring to business for the proactive measures to be taken in order to reduce the risks to the minimum level before they become problems. Furthermore, AI frees human biases and errors to improve accuracy in GRC processes. As AI will help compliance systems interpret regulatory requirements with precision and lower the chances of misinterpretation, organizations will automatically stay compliant with the legal standards.

Another big advantage of GRC solutions driven by AI is scalability. However, its complexity increases as more jurisdictions with different regulatory requirements are added to businesses operating internationally. For multinational organization navigating through the 'maze' of complex legal environment, the AI powered compliance platforms can handle the large volumes of regulatory data making it easier. Moreover, automation results in cost reduction by eliminating the demand for large compliance team and extensive manual risk assessment, enabling businesses to keep up compliance and cut operational expenditure.

However, there are many benefits of AI driven GRC strategy, but there are numerous challenges for organizations to implement it. The main issue is about data privacy and security. Large datasets such as sensitive business and customer information are used by AI driven compliance and risk management tools. To maintain ethical and legal standards, AI systems cannot be built without the adherence to data protection regulations including the General Data Protection Regulation (GDPR), the California Consumer Privacy Act (CCPA), etc.

Regulatory uncertainty is another challenge. The regulation of AI is evolving itself and businesses have to be informed about the new laws related to AI governance and accountability. There is also no standard regulation of AI in industries or countries, therefore making the compliance exercise complicated. Moreover, it is necessary for organizations to deal with algorithmic bias in AI implemented risk and compliance models. Training data biases can result in unfair risk assessments, wrong prediction of compliance and discriminatory decision making.

Integral to this complexity is the difficulty of integrating AI into a GRC program. However, many of the organizations work on legacy systems which are not built to be used with AI driven solutions. To use AI successfully in an enterprise environment without committing incomprehensible levels of dollars to technology upgrade, and changing billions of lines of data — and training employees to do the new things that AI enablers don't do — it needs significant investment in technology infrastructure and data infrastructure as well as compliance, risk management and governance programs. Additionally, organizations must fill the skill gaps by having compliance and risk management professionals skilled enough to understand the meaning of insight generated by AI and making the right decisions. It is believed that the role of AI technology in the future of GRC will increase further as this technology evolves. One important trend is that of explainable AI (XAI) in that it makes the AI driven compliance and risk management decision more transparent. The explain ability is needed to guarantee the trust, auditability and understandability of the AI recommendations to the regulators and stakeholders. AI in GRC will enable organizations to better achieve compliance requirements and increase stakeholder confidence.

A new is growing of integration of AI with blockchain technology for compliance tracking, and auditability. By employing blockchain, organizations can maintain verifiable records of the compliance activities, without the requirement of having centralized fiduciary or integrity custodians. According to the source, AI is being used to develop blockchain solutions that automate regulatory reporting, improve supply chain traceability as well as fraud detection machinery.

Also, AI powered regulatory sandboxes are expected to taste the success where businesses get a chance to test the AI powered compliance solution in a controlled space. It brings out the fact that regulatory bodies are increasingly experimenting AI driven sandboxes to check how AI regulations will impact the industry, and what should be the norms of compliance before making them compulsory industry wide. Moreover, the self-learning GRC systems that are driven by AI will continue to evolve, learning new risks, regulatory changes and industry changes without manual intervention.

GRC, through its art of understanding risks, compliance processes and devoting efforts for good governance, is undergoing transformation with the advent of AI – enhancing risk assessments with AI powered risk assessments, automating compliance processes with AI for faster and smooth execution of compliance processes and supporting data driven governance with AI. With the future of GRC shaping by AI, businesses need to be innovative and take control of the ethical governance of AI and human oversight. With AI/GRC being core to the thriving future, organizations can build a future that is resilient and compliant, by adopting AI-driven strategy for GRC strategies.

5.4.1 AI for intelligent regulatory compliance and auditing

The regulatory compliance and auditing are being changed by transforming AI based processes, which automate processes, increase accuracy and also make it possible for real time monitoring. Manual oversight in itself is a compliance and auditing framework that is time consuming, prone to human error and does not scale in size to handle large data volume. Compliance solutions powered by artificial intelligence and machine learning examine all regulatory prerequisites throughout several countries and understand an organization's right to know obligation so it can stay current with changing legal requirements. The purpose for these systems is for them to auto detect possible compliance violations, regulatory risks and generate real time alerts and at the same time allow businesses to take proactive corrective actions before the compliance issues escalate. It cuts down compliance checks on the manual side by increasing efficiency and accuracy of regulatory adherence (Murdock, 2021).

In the world of auditing, AI helps in risk assessment by process of analyzing entire dataset instead of resorting to small samples. By using machine learning models, auditors are alerted by financial transactions and operational records anomalies, inconsistencies and potential fraud to detect. The risk scoring systems are driven by AI, which prioritizes high risk transactions and allows the auditors to concentrate on the critical areas and try to streamline the process of audit. More importantly, AI enabled automated auditing tools can reduce the reporting errors by making sure the compliance documentation is according to regulatory standards. With a great use of AI, organizations can produce compliance reports, automated audits and audit trails using less human intervention, which increases transparency and accountability.

Cybersecurity in AI is utilized by Compliance Monitoring, which is AI driven and analyzes user behavior, network activity and financial transactions to identify suspicious patterns. In real time, these intelligent systems detect the unauthorized access, fraudulent activities and regulatory breaches for the organizations to handle the risks quickly. Virtual assistants and chatbots are also AI powered which helps the compliance teams in answering regulatory queries, guiding team member regarding the policy adherence and easier handling of compliance reporting. Additionally, combining AI with blockchain technology is making regulatory transactions more transparent with immutable record of such transactions, and thereby prepping auditing process by enhancing compliance transparency.

AI based audits and compliance is certainly a great idea, however like any other, these have their challenges, data privacy, algorithmic bias and a vague regulatory status. With models of AI demanding access to huge quantities of sensitive business and financial data to build themselves, they have to be guarded according to GDPR and CCAPA. AI algorithms must be fair and bias must be audited often to continually ensure the compliance assessments are not discriminatory. To continue with the regulatory bodies changing the AI governance framework, businesses will have to have ethical AI deployment, explain ability and accountability coupled with AI compliance and auditing systems based on trust in AI driven systems. Companies that are able to leverage AI to do compliance and auditing will get a competitive advantage in the competitive business environment by cutting operational costs, managing the risks and strictly following the regulations of the industry.

5.4.2 AI's impact on corporate governance and decision-making

It is with great force that all business processes and decision-making processes are being transformed by Artificial Intelligence (AI). New intelligence and agility is being brought to the world through use of vast datasets, through machine learning algorithms and predictive analytics, through the use of AI. AI systems contribute to the board of directors and top management abilities to monitor the organizational performance, to spot risks of misconduct or inefficiency at the early stage, and to ensure compliance with the increasingly complicated regulatory environment. Such systems are able to constantly process financial records, legal documents, operational data and even employee communications to look for anomalies or risks. Thus, governance is more proactive, data driven, and resilient to fast changing technological, economic and geopolitical change (Vemuri, 2020).

However, main role of present AI is to further strengthen transparency and accountability. AI automates the production of audit trails and real time reporting, giving stakeholders to look at the corporate activities and decisions with accurate and up to date information. It instills more trust among the investors, regulators and the general public. Additionally, AI tools can help foster the corporate ethics and sustainability practices by constantly monitoring environmental, social, and governance (ESG) indicators and notifying top leadership about the deviations from corporate responsible goals. It enables organizations to meet the regulatory expectation whilst also making fit operations with wider societal values.

On the matter of decision making, nothing comes close to the speed, precision and foresight that is AI. In this way, AI powered decision support systems can be used by senior management to simulate several strategic scenarios, to evaluate risk under different economic situations and to optimize resource allocation across departments or business units. One of these capabilities substantially reduces certainty which allows these organizations respond better to market volatility, customer behavior shift or supply chain disruption. For example, in investments, AI can process years of accounting, macroeconomic trends, competitive landscape in seconds and provide information to help make calculation decisions that are more confident.

On the other hand, AI helps to get rid of cognitive bias, and to be consistent in the decision making. These factors of traditional human decision making such as overconfidence, anchoring, or groupthink can influence human decision making. If designed and governed responsibly, AI is a neutral advisor that presents these kinds of data driven recommendations without a bias. But it should be understood that AI is not infallible. Its quality of outputs depends very much on the quality and diversity of data it is trained on. For this reason, AI generated insights still require human oversight in interpreting them, validating outcomes, and to ensure that the corresponding outcomes are aligned with the company's strategic vision, ethical standards and long-term business objectives.

Despite these benefits, the integration of AI into corporate governance and decision-making raises important challenges and ethical considerations. Important issues to take care of are things like data privacy, algorithmic bias, explain ability of AI decisions, and at the same time preventing human accountability from being eroded. For AI itself, organizations have to create clear governance frameworks —by defining roles, responsibilities, and rules on how it is to be applied. As an example, this includes being transparent, auditable and frequently tested for fairness and accuracy of AI systems. There should be members of the board who have the right digital literacy or digital acumen to understand and supervise AI initiatives.

Finally, AI is not just a technology, but a strategic enabler that is changing the very nature of the governance of corporations, how they choose to make decisions. Due to its speed of processing information, uncover and hidden insights, and Coders forecast future trends, it represents a competitive advantage in today's climate. But for that, AI must be responsibly integrated by organizations, not supplanting but complementing human judgment. The same cannot be said about any of the other high-tech inventions which have come before. If AI is properly guided through an appropriate balance of innovation, oversight, and ethical governance, it can function as a formidable engine for smarter, more transparent, and more responsible corporate leadership in the coming years.

5.5 Ethical AI Adoption and Responsible Implementation

In the age of intelligent systems, adoption of and responsible implementation with ethical AI are the two fundamental pillars to high-trust adoption, compliance and digital innovation. Considering the expanding role of artificial intelligence as it is deeply embedded into organization and business processes; products and services, ethical implications of deployment of AI become more critical as challenges. AI practices with a commitment to guard AI development and use in the respect of human rights and the promotion of fairness and minimization of harm are responsible. It demands intentional and conscious work in determining, and in turn obfuscating, the specific way in which AI systems booby trap data while also maintaining privacy and disclosing decision making. To adopt ethical AI, we also take up ensuring secured informed consent from the users collecting and processing data, and the ability for the humans to overwrite any automated decision as and when required (Dimple Patil, Jayesh Rane, Nitin Liladhar Rane, 2024).

From a governance view point, organizations have to have proper policies and oversight to steer the ethical evolution and use of AI technologies. It includes setting up interdisciplinary ethics committees, installing AI ethics into the corporate values, and laying out concrete responsibility structures that determine who is accountable for the AI impact. In addition, AI systems should be explainable as they should be able not only to meet regulatory requirements, but also to support user trust and encourage meaningful human-AI collaboration. It also requires the frequent auditing, assessing of impact and risk of the AI's use to ensure its continued alignment with emerging legal and social expectations of ethical AI.

Importantly, responsible AI is a cultural and strategic as much as a technical and legal issue. Since organizations comprise a set of people, they must develop a culture of ethical awareness in which the employees learn responsible AI practices and openly talk about the moral implications of emerging technologies. However, working alongside the external stakeholders like regulators, academic institutions as well as civil society groups will not only keep companies in the loop on the best practices but also about the alignment w district societal values. With this, the company creates the foundation for ethical AI to become a source of competitive advantage that not only strengthens reputation, but also stimulates innovation and puts the firm at the forefront of digital responsibility. As a result, ethical AI adoption is not simply remotely avoiding harming – it is about doing good with AI, using AI to help empower people, to facilitate more inclusive growth and to reach towards the more equitable and sustainable world possible.

5.5.1 Addressing AI bias and fairness in decision-making

Dealing with AI bias and fairness of decision making is an important problem in AI system development and deployment. When it comes to decisions of life or death, or for that matter even basic rights — hiring and credit, hiring and credit, healthcare, or the criminal justice system — it is fair to question whether using AI in such high stakes can potentially reinforce or even amplify the existing inequalities. Once the bias has entered the AI system, it can occur at multiple points through training data that is skewed, modeling assumptions that are flawed or representation of social groups that are out of balance. Being seen in these terms, if not controlled, can lead to unfair outcomes like discriminating against some particular individual or community as well as undermining public confidence in the use of AI technologies. To address such risks, fairness should be considered a key design principle and the datasets should be proactively audited for imbalances to prevent issues from occurring. It may include data re sampling, fairness constraints during algorithms training as well as bias mitigation strategies in algorithms evaluation (Schwartz et al., 2022).

Fairness in AI also is not a technical issue only but a socio ethical one. There are many possible definitions of fairness in various contexts, based on values and legal frameworks of a particular context as they would be equal opportunity, demographic parity, or individual fairness. Thus, fairness objectives together with socio-digital impact evaluation of AI systems are dependent on meaningful stakeholder engagement. Likewise, transparency is also important in issuing teams accountability, namely, the idea that organizations ought to present the trail of how AI model development, use of the data and the method of decision making. Another way of using explainable AI techniques is to help users better understand model outputs and find potential sources of bias in an AI system.

However, being fair in building AI is an important interdisciplinary work of building between data scientists, ethicists, lawyers, domain experts and affected communities. Fairness reviews and ethical oversight should have governance structures that apply throughout the AI's lifetime. Conflict of interest policies can also be developed, along with industry standards in support of fairness efforts, but corporate responsibility remains the foremost endeavor of business. Finishing bias and fostering fairness in AI is not just about compliance, it is about developing inclusive, appropriate technologies that are satisfactory to everyone in society and rebuild important ethical ideals that make trust, and progress, feasible.

5.5.2 Ensuring transparency and accountability in AI adoption

Responsible AI adoption is based on the principles of transparency and accountability to ensure the artificial intelligence systems work in a way that is understandable, traceable and in line with ethical, legal and other standards. When AI is ingrained so much into decision making in an organization, one of its often-opaque processes such as complex models like deep learning can be difficult to deal with. Having said that, transparency is in regards to making the inner workings of the AI systems, the data used to iteratively improve the systems to give the most accurate output, as well as the reconciliation for the use of the data and why you are making a decision in the first place, be visible and interpretable to concerned stakeholders. Without this clarity, it was not possible to judge whether AI decisions were fair, unbiased and accurate. Therefore, explain ability is of utmost importance in the AI systems that organizations need to develop, so that users, regulators, or affected individuals are able to understand how this outcome is derived (Masís, 2021).

Data provenance and model interpretability is one of the main aspects of transparency. Documentation of data sources, preprocessing steps, labeling practices and assumptions of model development should be kept by organizations. It enables reproduction and insight into how certain variables affect AI predictions. Further, this provides opportunities to build models where the interpretable nature is built in as is employed in explainable AI techniques. For example, in healthcare, if a person suggested by an AI model for a treatment plan without any clear benefits to the patient based on clinical data, it would not work. It makes it possible for clinicians to verify human AI collaboration instead of blind reliance.

The accountability side, however, requires setting up mechanisms for oversight, ownership and redress in case certain effects follow from use of AI systems. The design, deployment, and monitoring of AI systems must be defined by organizations in terms of who is responsible. One way this could be included is creating AI ethics boards of their own or holding senior leadership accountable to make sure the use of AI meets the company's values and statutory obligations. Additionally, there must be clear paths through which an AI decision can cause harm—the system must not only fail to offer these explanations, compel redress, or be an avenue for investigation, it must also stand between these explanations, redress, or investigation. Yet business now has to consider the legal and regulatory frameworks that increasingly are mandating such accountability and therefore has no choice but to put governance structures in place proactively.

Auditing, ongoing as well as impact assessments, also need to be carried out. These tools let organizations monitor the performance of a system, understand their risk and ensure that the AI systems operate as they are expected to over time and after deployment. AI development process needs to be algorithmic audited, fairness evaluated, and bias detection incorporated either into the AI lifecycle (from design through to deployment and beyond). In addition to that, adopting AI transparency standards, for instance model cards, data sheets and audit trails, allow stakeholders to access the information they need to evaluate integrity of the system. Such efforts should be supported by stakeholder engagement to gather feedback from users, customers, and communities impacted by the organization.

More than anything, being transparent and accountable with AI systems is not a risk mitigation exercise—although it mitigates risk—but an exercise of building the public trust in AI systems. Fairness of AI increases the chances of its adoption and support by stakeholders if believed to be working fairly with controls and can be understood. Embedding transparency and accountability into your AI strategy's strategy shows that the company is seeking ethical innovation and sustainability in its digital transformation. These principles give users power, protect individuals, and provide the public confidence to build the responsible use of these technologies, enabling AI to be used for the good of all.

Multiple Choice Questions (MCQs)

1. What is the primary goal of integrating AI in future-proofing enterprises?

A. Reducing employee count

B. Automating only marketing functions

C. Enhancing adaptability and resilience in changing environments

D. Eliminating human decision-making

2. Which of the following is an emerging AI trend in enterprise resource planning (ERP)?

A. Manual ledger reconciliation

B. Rule-based accounting

C. Predictive analytics and intelligent automation

D. Employee time-tracking via spreadsheets

3. How does AI enhance cybersecurity in modern enterprises?

A. By creating manual firewalls

B. By identifying and responding to threats in real time using intelligent algorithms

C. By deactivating user accounts periodically

D. By relying on physical security tools

4. What is one-way AI contributing to risk-aware business strategies?

A. Reducing the need for data analysis

B. Making decisions without data

C. Providing predictive insights for proactive risk management

D. Outsourcing risk evaluation to third parties

5. How does AI synergize with blockchain for enhanced security?

A. By encrypting data manually

B. By replacing blockchain functionality

C. By ensuring real-time anomaly detection and immutable audit trails

D. By reducing data transparency

6. In AI-powered IoT systems, what key role does AI play in risk management?

A. Disabling IoT device functions

B. Ignoring sensor data

C. Detecting anomalies and predicting system failures

D. Monitoring only physical infrastructure

7. What is the benefit of using AI in Governance, Risk, and Compliance (GRC) strategies?

A. Generating fixed reports monthly

B. Enforcing rigid manual compliance processes

C. Automating compliance monitoring and providing intelligent auditing

D. Avoiding audits

8. In what way does AI influence corporate governance?

A. By centralizing all decisions to AI

B. By replacing boards of directors

C. By offering data-driven decision support and improving transparency

D. By focusing only on finance

9. Why is addressing AI bias important in responsible implementation?

A. To limit algorithmic capabilities

B. To favor specific user groups

C. To ensure fair and equitable outcomes for all users

D. To simplify data labeling

10. How can organizations ensure transparency and accountability in AI systems?

A. By keeping algorithms secret

B. By using black-box models without documentation

C. By applying explainable AI, auditing systems, and defining responsibility

D. By allowing AI to self-regulate

Answers

1	2	3	4	5	6	7	8	9	10
C	C	B	C	C	C	C	C	C	C

Bibliography

Access, O., Manoharan, A., & Sarker, M. (2024). Revolutionizing Cybersecurity: Unleashing the Power of Artificial Intelligence and Machine Learning for Next-Generation Threat Detection. *International Research Journal of Modernization in Engineering Technology and Science*, *12*, 2151–2164. https://doi.org/10.56726/irjmets32644

Agarwal, P., Topno, P. G., & Chandan, M. (2020). Role of AI in Enhancing IoT Security. *International Journal of Psychosocial Rehabilitation*, *27* (3), 10123–10126. https://doi.org/10.61841/v24i1/400335

Ahmed, N. A. (2023). *NLP techniques used for compliance checks.*

airbyte. (2024). *What Is AI Workflow Automation: Tools, Use Cases, & More*. Airbyte.

Al-Quayed, F., Ahmad, Z., & Humayun, M. (2024). A Situation Based Predictive Approach for Cybersecurity Intrusion Detection and Prevention Using Machine Learning and Deep Learning Algorithms in Wireless Sensor Networks of Industry 4.0. *IEEE Access*, *PP*, 1. https://doi.org/10.1109/ACCESS.2024.3372187

Ambassadors, S. (2024). *AI-Driven Decision Support Systems*. Medium.

Amblard-Ladurantie, C. (2024). *How Artificial Intelligence Can Be Used in Compliance*. MEGA.

Anamika Pandey, Balamurugan Balusamy, N. C. (2023). *Disruptive Artificial Intelligence and Sustainable Human Resource Management Impacts and Innovations – The Future of HR.*

arkoselabs. (2024). *What is fraud detection?* Arkoselabs.

Artmaster. (2024). *The Role of AI in Disaster Management: Empowering Resilience*. Artificial Paintings.

Azeem, M., Khan, D., Iftikhar, S., & Bawazeer, S. (2024). Heliyon Analyzing and comparing the effectiveness of malware detection : A study of machine learning approaches. *Heliyon*, *10* (1), e23574. https://doi.org/10.1016/j.heliyon.2023.e23574

Bhumichai, D., Smiliotopoulos, C., Benton, R., Kambourakis, G., & Damopoulos, D. (2024). The Convergence of Artificial Intelligence and Blockchain: The State of Play and the Road Ahead. *Information (Switzerland)*, *15* (5), 1–32. https://doi.org/10.3390/info15050268

BluEntTech. (2023). Fraud Data Analytics: How to Detect and Prevent Fraud Using Data. *BluEntTech*.

Bojan Kolosnjaji, Huang Xiao, Peng Xu, A. Z. (2024). *Artificial Intelligence for Cybersecurity Develop AI Approaches to Solve Cybersecurity Problems in Your Organization*.

Briefs, S., & Business, I. N. (2018). *Arti cial Intelligence for Business 123*.

By, Anthony TAEditedrantino, P. (2008). C Ompliance H Andbook. *Management*, 957.

Choudhuri, S. S. (2024). *AI IN ERP AND SUPPLY CHAIN MANAGEMENT*.

Coplin, D. (2024). *The Rise of the Humans: How to outsmart the digital deluge How to outsmart the digital deluge*.

Das, R. (2024). *A Reference Manual for Data Privacy Laws and Cyber Frameworks*.

Data, B., Asia, E., & States, U. (2016). Managing Cyber Risk in the Financial Sector. In *Managing Cyber Risk in the Financial Sector*. https://doi.org/10.4324/9781315675930

Dhieb, N., Ghazzai, H., Besbes, H., & Massoud, Y. (2020). Scalable and Secure Architecture for Distributed IoT Systems. *2020 IEEE Technology & Engineering Management Conference (TEMSCON)*, 1–6. https://doi.org/10.1109/TEMSCON47658.2020.9140108

Dimple Patil, Jayesh Rane, Nitin Liladhar Rane, P. D. (2024). *Trustworthy Artificial Intelligence in Industry and Society.*

Ellen Cibula. (2023). *How AI Is Transforming Risk Assessment in Finance.* Alibabacloud.

Elliott C. Kulakowski, L. U. C. (2006). *Research Administration and Management.*

fastercapital. (2024). *Financial Behavior: How to Understand and Influence Financial Behavior for Your Business.* Fastercapital.

Ferdin Joe John Joseph, Karthikeyan Chinnusamy, P. Paramasivan, R. Regin, S. S. R. (2024). *Explainable AI Applications for Human Behavior Analysis.*

Gabija Stankevičiūtė. (2023). *AML Automation: Streamlined Compliance 101 for Businesses.* Idenfy.

Gilboa, I., & Schmeidler, D. (2015). a C Ognitive M Odel. *International Journal of Emerging Technologies in Learning (IJET), 10* (3), 62–65.

Gill, N. singh. (2024). *Deep Learning in Cybersecurity: Threat Detection and Defense.* Xenonstack.

Greengard, S. (2021). *The Internet of Things, Revised and Updated Edition.*

Harapan, P., Publishing, E., Collection, B. A., An, U., Leung, D., Accounting, I., Sociology, T., Reporting, F., & Account, A. (2020). *A Gower Book.*

He, H. (2011). *Self-Adaptive Systems for Machine Intelligence.*

HighRadius. (2024). *Complete Guide to Data Anomaly Detection in Financial Transactions.* Highradius.

Horowitz, M. C., Allen, G. C., Saravalle, E., Cho, A., Frederick, K., & Scharre, P. (2018). UNCLASSIFIED // FOR OFFICIAL USE ONLY Artificial Intelligence and International Security. *The Artificial Intelligence and International Security Series, July,* 1–28.

INTERROLL. (2024). *AI Warehouse Automation: The Future of Logistics Efficiency.* INTERROLL.

Intersystems. (2023). *Trust and Responsibility with AI*. Intersystems.

Jagreet Kaur, N. S. G. (2019). *Artificial Intelligence and Deep Learning for Decision Makers A Growth Hacker's Guide to Cutting Edge Technologies*.

Karim, R., Galar, D., & Kumar, U. (2023). *AI Factory*. CRC Press. https://doi.org/10.1201/9781003208686

Koushik, P. (2024). *Supply Chain Synergy Integrating AI and ML for Optimal Order Management*.

Kurt J. Engemann, D. M. H. (2014). *Business Continuity and Risk Management Essentials of Organizational Resilience*.

Lab, S. (2023). *Artificial Intelligence in Risk Management*. Lab, SuDu.

Lewis, T. G. (2006). Critical Infrastructure Protection in Homeland Security: Defending a Networked Nation. *Critical Infrastructure Protection in Homeland Security: Defending a Networked Nation*, 1–474. https://doi.org/10.1002/0471789542

Linh Chu Dieu. (2024). *AI in Anti-Money Laundering: Revolutionizing Compliance in the Fight Against Financial Crime*. Smartdev.

LINK, L. (2024). *AI in Surveillance Systems: How to Empower Security Solutions with AI*. LITS LINK.

Llauradó, J. M., Pujol, F. A., Tomás, D., Visvizi, A., & Pujol, M. (2023). Study of image sensors for enhanced face recognition at a distance in the Smart City context. *Scientific Reports*, *13* (1), 1–13. https://doi.org/10.1038/s41598-023-40110-y

Mandl, C. (2023). *Procurement Analytics Data-Driven Decision-Making in Procurement and Supply Management*.

Masís, S. (2021). Interpretable Machine Learning with Python Build explainable, fair, and robust high-performance models with hands-on, real-world examples. *Packt Publishing*, 737.

Matthews, P., & Greenspan, S. (2020). Automation and Collaborative Robotics: A Guide to the Future of Work. In *Automation and*

Collaborative Robotics: A Guide to the Future of Work. https://doi.org/10.1007/978-1-4842-5964-1

Maxim Dupont. (2023). *Introduction To AI In Finance*. Labelvisor.

Meris, R. (2025). *Leveraging AI in Regulatory Reporting*. Safety Culture.

Moallem, A. (2021). *Smart and Intelligent Systems The Human Elements in Artificial Intelligence, Robotics, and Cybersecurity*.

Murdock, H. (2021). *Operational Auditing*. CRC Press. https://doi.org/10.1201/9781003096931

N.A., S. A., Veluri, R. K., N.A., B. R. T. B., & Rajesh, C. (2025). Deep learning-based anomaly detection in video surveillance. *International Journal of Electronic Security and Digital Forensics*, 17 (4). https://doi.org/10.1504/ijesdf.2025.10064692

Nikhil Saini. (2025). *The Role of Predictive Analytics in Fraud Prevention*. Hashstudioz.

Omar Santos, P. R. (2024). *Beyond the Algorithm AI, Security, Privacy, and Ethics*.

Paltrinieria, N., Comfortb, L., & Reniersc, G. (2019). Learning about risk: Machine learning for risk assessment. *Science Direct*.

Paras Kela. (2024). *Artificial Intelligence in ERP: How it is Shaping the Future of Businesses?* Softwaresuggest.

Parisi, A. (2019). *Hands-On Artificial Intelligence for Cybersecurity*.

Paul Daugherty, H. J. W. (2023). *Consequently, the use or augmentation of AI driven technologies are replacing or supplementing traditional warehousing processes, which typically rely on manual labour and legacy systems.*

Puplampu, R. (2024). *What Everyone Should Know About the Rise of AI AI Transparency, Privacy, and Ethics Best Practices*.

Rieple, A., DeFillippi, R., & Schreiber, D. (2023). *Transformational Innovation in the Creative and Cultural Industries*. Routledge. https://doi.org/10.4324/9781003207542

Robert Baldwin, Martin Cave, M. L. (2011). *Understanding Regulation Theory, Strategy, and Practice.*

Rohit Rao. (2024). *Continuous Monitoring: What It Is, Benefits, Types & More*. Zluri.

Schildt, H. (2020). *The Data Imperative How Digitalization Is Reshaping Management, Organizing, and Work.*

Schwartz, R., Vassilev, A., Greene, K., Perine, L., Burt, A., Hall, P., & Greene, K. (2022). NIST special publication 1270. Towards a standard for identifying and managing bias in artificial intelligence. *NIST Special* ..., 1–86.

Science, E. (2024). *Computational Intelligence Techniques for Sustainable Supply Chain Management.*

Seker, E. (2020). *Is the Future of Cyber Security in the Hands of Artificial Intelligence (AI)? — 1*. TDS Archive.

sensfrx. (2023). *The Use of Behavioral Analytics In Fraud Detection*. Sensfrx.

Shemshaki, M. (2024). *The Benefits of Using Artificial Intelligence for Business Success Strategies for Innovation, Efficiency, and Growth.*

SIEVO. (2024). *AI in Procurement*. SIEVO.

Skene, K. R. (2019). *Artificial Intelligence and the Environmental Crisis Can Technology Really Save the World?*

Spedding, L., & Rose, A. (2018). *Business Risk Management Handbook*. 17–63.

splendapp. (2024). *Biometric Authentication: It's Meaning And Benefits*. Splendapp.

Takyar, A. (2024). *AI in demand forecasting: Use cases, benefits, solution and implementation*. LeewayHertz.

TARLOGIC. (2023). *What are the AI security risks?* TARLOGIC.

Ummah, M. S. (2019). Disruptions, challenges, and crises have become an uncertain reality for organizations in the world that we live in. *Sustainability (Switzerland)*, *11* (1), 1–14.

Usha, G. (2024). *Utilizing AI and Machine Learning for Natural Disaster Management. September.* https://doi.org/10.4018/979-8-3693-3362-4.ch017

Vemuri, V. K. (2020). The AI advantage: how to put the artificial intelligence revolution to work. *Journal of Information Technology Case and Application Research*, *22* (1), 70–72. https://doi.org/10.1080/15228053.2020.1756084

Victoria Shutenko. (2024). *AI Anomaly Detection: Best Tools And Use Cases*. Techmagic.

Von Braun, J., Archer, M. S., Reichberg, G. M., & Sorondo, M. S. (2021). Robotics, AI, and humanity: Science, ethics, and policy. In *Robotics, AI, and Humanity: Science, Ethics, and Policy*. https://doi.org/10.1007/9783030541736

Walters, R., & Novak, M. (2021). Cyber Security, Artificial Intelligence, Data Protection & the Law. *Cyber Security, Artificial Intelligence, Data Protection & the Law*, 1–458. https://doi.org/10.1007/978-981-16-1665-5

Yan, W. Q. (2019). *Introduction to Intelligent Surveillance*.

About the Authors

My name is **Ashish Joon,** and I am based in Pensacola, Florida, USA. I currently work as a Software Developer and IT Professional, with extensive experience in .NET development, enterprise application design, and systems integration. Over the years, I have led and contributed to mission-critical software projects across diverse industries, with a strong focus on backend development, process automation, and scalable solution architecture.

My technical skill set is rooted in the Microsoft technology stack, with hands-on experience in C#, ASP.NET, Angular, SQL Server, Oracle, and related frameworks. I have successfully delivered a range of enterprise applications by combining engineering precision with agile methodologies, ensuring both technical excellence and business alignment.

In addition to my professional work, I am actively involved in the academic and research community, particularly in the field of Artificial Intelligence and Machine Learning (AIML). I have published multiple research papers in IEEE conferences, where my work has contributed to advancing AI-driven systems and intelligent automation. I also serve as a reviewer for IEEE conferences and frequently participate as a guest lecturer, sharing insights on emerging technologies and real-world implementation strategies.

My dedication to mentoring and community engagement includes serving as a judge for coding competitions and hackathons, where I evaluate technical creativity, code quality, and problem-solving approaches.

I hold a Master of Science in Computer Science from the New York Institute of Technology (NYIT) and a Bachelor of Engineering in Computer Science from OIST Bhopal, India. I remain committed to continuous learning and actively follow advancements in AI, cloud infrastructure, and software engineering practices.

Outside of work, I enjoy swimming, fitness, and exploring new technologies. I believe in leading with integrity, mentoring with purpose, and building solutions that create real impact.

My name is **Harshit Kohli** and I am based out of Raleigh, North Carolina USA. I am currently working with a leading tech company, bringing over 14 years of rich experience in IT infrastructure management and solution design. My career has been marked by a strong track record in enhancing cloud strategies specifically tailored for healthcare clients, implementing robust analytics solutions, and effectively managing workloads related to Artificial Intelligence. I have received recognition for my ability to train and mentor teams, drive business process improvements, and conduct thorough architecture reviews, all of which play a crucial role in ensuring client success in a competitive landscape. Currently, I am pursuing a Doctorate in Artificial Intelligence at the University of the Cumberlands in the USA, complementing my academic background that includes a Master of Science in Information Technology and a Bachelor of Technology in Electronics and Instrumentation from India. My extensive experience encompasses collaboration with numerous clients on various Analytics and Artificial Intelligence use cases, where I have consistently demonstrated exceptional efforts and achievements. Additionally, I have had the honor of serving as a keynote speaker at several international conferences focused on Analytics, Computing, and Artificial Intelligence. My involvement extends to reviewing research papers for multiple conferences and chairing sessions to ensure their successful organization. Furthermore, I have acted as a judge for various hackathons and prestigious awards, including Globee and Business Intelligence. Overall, my expertise in Data Analytics and Artificial Intelligence not only enhances my professional capabilities but also deepens my understanding of the field.

www.ingramcontent.com/pod-product-compliance
Ingram Content Group UK Ltd.
Pitfield, Milton Keynes, MK11 3LW, UK
UKHW060117010826
14089UKWH00004B/37

9 798899 844003